Paradox

www.**transworldbooks**.co.uk

Paradox

The Nine Greatest Enigmas in Science

Professor Jim Al-Khalili

BANTAM PRESS

LONDON · TORONTO · SYDNEY · AUCKLAND · JOHANNESBURG

TRANSWORLD PUBLISHERS
61–63 Uxbridge Road, London W5 5SA
A Random House Group Company
www.transworldbooks.co.uk

First published in Great Britain
in 2012 by Bantam Press
an imprint of Transworld Publishers

A CIP catalogue record for this book
is available from the British Library.

ISBN 9780593069295 (cased)
9780593069301 (tpb)

Addresses for Random House Group Ltd companies outside the UK
can be found at: www.randomhouse.co.uk
The Random House Group Ltd Reg. No. 954009

The Random House Group Limited supports the Forest Stewardship Council (FSC®), the
leading international forest-certification organization. Our books carrying the FSC label are
printed on FSC®-certified paper. FSC is the only forest-certification scheme endorsed by
the leading environmental organizations, including Greenpeace. Our paper procurement
policy can be found at www.randomhouse.co.uk/environment.

Typeset in 11/16 pt Berkeley Book by
Falcon Oast Graphic Art Ltd.
Printed and bound by CPI Group (UK) Ltd, Croydon, CR0 4YY

2 4 6 8 10 9 7 5 3 1

To Julie, David and Kate

Contents

Acknowledgements

I HAVE HAD tremendous fun writing this book. Much of its content has slowly accumulated over the course of my career teaching undergraduate physics, where I have used many of the paradoxes discussed and dissected in the following chapters as examples in my lectures to highlight and explain difficult concepts in relativity and quantum theory. Having said that, I must thank several people for their advice and support over the past year. My literary agent, Patrick Walsh, has, as always, provided unstinting friendly encouragement, as has my editor at Transworld, Simon Thorogood, and Vanessa Mobley at Crown. I am also hugely indebted to my copy-editor Gillian Somerscales for her many helpful comments, corrections and persistence in getting me to make my explanations as clear as possible. I would also like to thank the many hundreds of undergraduate students whom I have taught over the years at the University of Surrey for 'keeping me honest' when it comes to the subtler aspects of modern physics. Last but not least, I wish to thank my wife, Julie, for her unflagging support and encouragement in everything I do.

Preface

PARADOXES COME IN all shapes and sizes. Some are straightforward paradoxes of logic with little potential for investigation, while others sit atop icebergs of entire scientific disciplines. Many can be resolved by careful consideration of their underlying assumptions, one or more of which may be faulty. These, strictly speaking, should not be referred to as paradoxes at all, since once a puzzle is solved it ceases to be a paradox.

A true paradox is a statement that leads to a circular and self-contradictory argument, or describes a situation that is logically impossible. But the word 'paradox' does tend to be used more broadly to include what I prefer to call 'perceived paradoxes'. For such puzzles there is a way out. It may be that the paradox has hidden within it a trick or sleight of hand that deliberately misleads the listener or reader. Once the trick is uncovered, the contradiction or logical absurdity disappears. Another type of perceived paradox is one in which the statement and the conclusions, while initially sounding absurd or at the very least counterintuitive, turn out on more careful consideration not to be so, even if the result remains somewhat surprising.

And then there is the category of paradoxes in physics. All of these – well, nearly all – can be resolved with a little bit of fundamental scientific knowledge; and these are the ones that I focus on in this book.

So let us first take a brief look at a true logical paradox, just so that it is clear what I am *not* going to be talking about. This is a statement that is constructed in such a way that there really is no way out of the loop.

Take the simple assertion: 'This statement is a lie.' On first reading, I imagine the words themselves will seem straightforward enough. Think about their meaning, however, and the logical paradox will become evident as you work carefully through the statement's implications. Can five simple words really give you a headache? If so, I would argue that it is a fun sort of headache, which is perhaps itself a paradox, and one that you will no doubt feel sadistically obliged to pass on to family and friends.

You see, 'This statement is a lie' is telling you that in announcing itself to be a lie it must itself be a lie, and so it is not a lie – in which case it is true, which is to say that it really is a lie, which means it is not a lie, and so on in an infinite loop.

There are many such paradoxes. This book is not about them.

This book is instead about my own personal favourite puzzles and conundrums in science, all of which have famously been referred to as paradoxes, but which turn out not to be when considered carefully and viewed from the right angle. While powerfully counterintuitive when first described, they always turn out to be missing some subtle consideration of the physics which, when taken into account, knocks out one of the pillars on which the paradox is built and brings the whole edifice toppling down. Despite having been resolved, many of them continue to be referred to as paradoxes, partly owing to the notoriety they gained when first articulated (before we had figured out where we were going wrong) and partly because, so presented, they are useful tools in helping scientists get their heads round some rather complex concepts. Oh, and because they are such delicious fun to explore.

Many of the puzzles we will look at do indeed seem at first to be true paradoxes rather than just perceived ones. That is the point. Take a simple version of the famous time travel paradox: what if you were to go back into the past using a time machine and kill your younger self? What happens to the killer you? Do you pop out of existence because you stopped yourself from growing older? If so, and you never did reach the age at which you became a murderous time traveller, then who killed the younger you? The older you has the perfect alibi: you never even existed! So if you did not survive to travel back in time and kill your younger self, then you do not kill your younger self and so you do survive to grow older and travel back in time and kill yourself, so you do that, so you don't survive, and so on. This appears to be the perfect logical paradox. And yet physicists have not yet ruled out the possibility, certainly in theory, of time travel. So how can we extricate ourselves from this paradoxical loop? I will discuss this particular paradox in Chapter 7.

Not all perceived paradoxes require a scientific background to make sense of them. To demonstrate this, I have given over the first chapter to a handful of such perceived paradoxes that can be resolved with commonsense logic. What do I mean by this? Well, consider a simple statistical paradox in which it is quite possible to draw the wrong conclusion from a basic correlation: it is known that towns with larger numbers of churches generally have higher crime rates. This is somewhat paradoxical, unless you believe that churches are breeding grounds for lawlessness and crime – which, whatever your religious and moral views, seems pretty unlikely. But the resolution is straightforward. Both a higher number of churches and a higher absolute level of crime are the natural results of a larger population. Just because A leads to B and A leads to C does not mean that B leads to C or vice versa.

Here is another example of a simple brainteaser that sounds paradoxical when first stated, but whose paradoxical nature dissolves away once it has been explained. It was recounted to me a few years ago by a Scottish professor of physics who is a colleague and close friend of mine. He claims that 'every Scotsman who travels south to England raises the average IQ of both countries'. The point is this: since all Scotsmen claim to be smarter than all Englishmen, then any one of them would enhance the average IQ of England by living there; however, to leave Scotland is such a foolish act that only those less intelligent among them would do so, leaving the average IQ of those remaining slightly higher. So you see, at first glance it sounds like a paradoxical statement, yet with simple logical reasoning it can be resolved beautifully – if not convincingly for the English, of course.

Once we have had some fun in Chapter 1 with a few well-known paradoxes that can be resolved without any science, we will move on to my nine chosen paradoxes in physics. After stating each one, I will lay it bare and explain how it evaporates away to reveal the underlying logic that shows its fallacy, or why it is not really an issue at all. They are all fun because they have some intellectual meat, and because *there is a way out*. You just need to know where to look, where to find the Achilles heel that can be exploited with careful prodding and a better understanding of the science, until the paradox is a paradox no more.

The names of some of these paradoxes will be familiar. Take the Paradox of Schrödinger's Cat, for instance, in which an unfortunate feline is locked into a sealed box and is simultaneously both dead and alive until we open the box. Less familiar, perhaps, but still known to some, is Maxwell's demon, the mythical entity that presides over another sealed box and which is seemingly able to violate that most sacred of commandments in science, the Second Law of Thermodynamics – forcing the contents of the box to un-mix and become ordered. To understand such paradoxes, and their resolution, it is necessary to grasp some background science; and so I have set myself the challenge of getting these scientific concepts across with as little fuss as possible, so that you can appreciate and enjoy the implications without any in-depth knowledge of calculus, thermodynamics or quantum mechanics.

I have plucked several of the other paradoxes in this book from the undergraduate course on relativity that I have taught for the past fourteen years. Einstein's ideas on space and time, for instance, provide fertile ground for logical brainteasers such as the Pole in the Barn Paradox, the Paradox of the Twins and the Grandfather Paradox. Others, such as those involving the cat and the demon, have, in the eyes of some, yet to be satisfactorily laid to rest.

When choosing my greatest enigmas in physics, I have not homed in on the biggest unsolved problems – for example, what dark matter and dark energy, which between them make up 95 per cent of the contents of our universe, are made of, or what, if anything, was there before the Big Bang. These are incredibly difficult and profound questions to which science has yet to find answers. Some, like the nature of dark matter, that mysterious stuff that makes up most of the mass of galaxies, may well be answered in the near future if particle accelerators like the Large Hadron Collider in Geneva continue to make new and exciting discoveries; others, like an accurate description of a time before the Big Bang, may remain unanswered for ever.

What I have aimed to do is make a sensible and broad selection. All the paradoxes I discuss in the following chapters deal with deep questions about the nature of time and space and the properties of the Universe on the very largest and smallest of scales. Some are predictions of theories that sound very strange on first encounter, but which become intelligible once the ideas behind the theory are explored carefully. Let's see if we can't lay them all to rest, and along the way give you, dear reader, some mind-expanding fun.

ONE

The Game Show Paradox

Simple probabilities that can really blow your mind

BEFORE I GET stuck into the physics, I thought I would lead you in gently with a few simple, entertaining and frustrating puzzles as a warm-up. In common with the rest of the collection in this book, none of these are real paradoxes at all; they just need to be unpicked carefully. But unlike those coming later, which will require an understanding of the underlying physics, the paradoxes in this chapter are a collection of logical brainteasers that can be resolved without any scientific background at all. The last and most delicious of the group, known as the Monty Hall Paradox, is so utterly baffling that I will invest considerable effort in analysing it in several different ways so you can choose which particular solution you prefer.

All the puzzles in this chapter fall into one of two categories with the exotic-sounding names of 'veridical' and 'falsidical' paradoxes. A veridical paradox is one leading to a conclusion that is counter-intuitive because it goes against common sense, and yet can be shown to be true using careful, often deceptively simple, logic. In fact, with these the fun lies in trying to find the most convincing way of demonstrating that it is true, despite that lingering uncomfortable feeling that there has to be a catch somewhere. Both the Birthday

Paradox, which I will discuss shortly, and the Monty Hall Paradox are in this category.

A falsidical paradox, on the other hand, starts off perfectly sensibly, yet somehow ends up with an absurd result. However, in this case the apparently absurd result is indeed false, thanks to some subtly misleading or erroneous step in the proof.

Examples of falsidical paradoxes are the mathematical tricks that, by following a few steps of algebra, 'prove' something like $2 = 1$. No amount of logic or philosophizing should convince you that this can be true. I won't go into any of these here, mainly since I don't really want to be hitting you with algebra just in case you don't love it as much as I do. Suffice it to say that the calculation leading to the 'solution' usually involves a step in which a quantity is divided by zero – something any self-respecting mathematician knows to avoid at all costs. Instead, I will focus on a few problems that you can appreciate with only the bare minimum of mathematical ability. I'll begin with two great falsidical paradoxes: the Riddle of the Missing Dollar and Bertrand's Box Paradox.

The Riddle of the Missing Dollar

This is a brilliant puzzle that I used a few years ago when I was a guest on a TV quiz show called *Mind Games* – not that I am claiming to have been the first to come up with it, of course. The premise of the show was that each week the guests would compete against each other to solve puzzles set by the host of the show, the mathematician Marcus du Sautoy. In addition, each was expected to bring along their favourite brainteasers to try to bamboozle the other team.

Here is how it goes:

Three travellers check into a hotel for the night. The young man at the reception desk charges them $30 for a room with three beds in it. They agree to split the price of the room equally, each of them paying

$10. They take the key and head up to the room to settle in. After a few minutes the receptionist realizes he has made a mistake. The hotel has a special offer on all week and he should only have charged them $25 for the room. So as not to get into trouble with his manager, he quickly takes five dollar bills from the till and rushes up to rectify his error. On the way to the room he realizes that he cannot split the five dollars equally between the three men, so he decides to give each of them one dollar and keep two for himself. That way, he argues, everyone is happy. Here, then, is the problem we are left with: each of the three friends will have contributed $9 towards the room. That makes $27 that the hotel has made, and the receptionist has a further $2, which makes $29. What has happened to the last dollar out of the original $30?

You may be able to see the solution to this straight away; I certainly didn't when I first heard it. So I will let you think about it a little before you read on.

Have you worked it out? You see, this puzzle only sounds paradoxical because of the misleading way it is stated. The error in the reasoning is that I added the $27 dollars to the $2 taken by the receptionist – and there is no reason to do that, because there is no longer a total of $30 that needs to be accounted for. The receptionist's $2 should be *subtracted* from the $27 paid by the friends, leaving $25, which is the amount in the till.

Bertrand's Box Paradox

My second example of a falsidical paradox is credited to the nineteenth-century French mathematician Joseph Bertrand. (It is not his most famous paradox, which is rather more mathematically technical.)

You have three boxes, each containing two coins; each box is divided into two halves by a partition, with a coin in each half. Each side can be opened separately to see the coin inside (that is, without allowing you to see the other coin). One box contains two gold coins (we will call this one GG), the second (which we will call SS) contains two silver coins and the third (GS) contains one of each. What is the probability of picking the box with the gold and silver coins in it? The answer is simple, of course: one in three. That is not the puzzle.

Figure 1.1. Bertrand's boxes

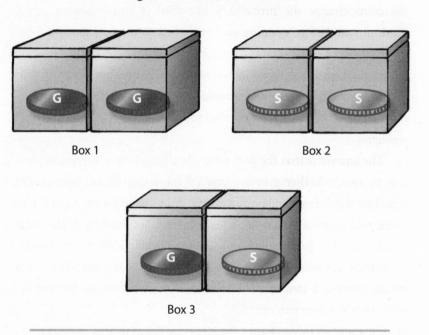

Box 1   Box 2

Box 3

Now, pick a box at random. What if you open one of its lids and find a gold coin inside? What now are the chances of this box being the GS one? Well, since in finding a gold coin you know this cannot be the SS box, you must rule that option out and be left with two choices: either it is the GG box or it is the GS box. Hence the probability of its being the GS box is one in two, right?

Had you opened the lid to find a silver coin instead, then you could now rule out the GG option; so you are left with SS or GS, so the probability that this is the GS box is still one in two.

Since you must find either a gold or a silver coin when you open the lid of the chosen box, and since there are three coins of each kind in total, giving you an equal chance of finding either, there is therefore a one in two probability that you have found the GS box whatever coin you find. Thus, after taking a peek inside one half of your chosen box the overall probability that it is the GS box must change from one in three, as it was at the start, to one in two. But how can seeing one of the coins change the probability like this? If you choose a box at random and, before opening one of its lids, you know that there is a one in three chance that it is the GS one, then how, by seeing one of the coins inside, and gaining no information at all from this, *since you know you are certain to find either a gold or a silver coin anyway*, does the probability switch from one in three to one in two? Where are we going wrong?

The answer is that the probability is always one in three and never one in two, whether you see one of the coins in the box or not. Consider the case when you find a gold coin inside your box. There are three gold coins in total – let us call them G1, G2 and G3, and let us say that the GG box contains coins G1 and G2, while G3 is the gold coin in the GS box. If you open one of the boxes and find a gold coin inside, there is a two in three chance of your having picked the GG box, since the coin you are looking at could be either G1 or G2. There is only a one in three chance that it is the G3 coin and therefore that the box you have picked is the GS one.

The Birthday Paradox

This is one of the best-known veridical paradoxes. Unlike the last two examples, there is no trick here, no error in the reasoning or sleight-of-hand in the telling. Whether you are convinced by the solution or not, I must stress that it is perfectly correct and consistent, both mathematically and logically. In a way, this frustration makes the paradox all the more fun.

Here is how it is stated:

How many people would you say there would have to be in a room for the chances of any two of them sharing a birthday to be better than fifty-fifty – that is, for it to be more likely than not that any two share a birthday?

Let us first apply a little naïve common sense (which of course is going to turn out to be wrong). Since there are 365 days in a year, imagine there is a lecture hall with 365 empty seats. One hundred students enter the hall and each of them takes a seat at random. Some friends may wish to sit next to each other, a few prefer the anonymity of the back row so they can fall asleep undetected, while the more studious prefer to be closer to the front. But it does not matter how they distribute themselves; the fact remains that more than two-thirds of the seats will remain empty. Of course, no student will sit on a seat already occupied, but we sort of feel that the chance of any two students wanting the same seat is pretty slim, given how much space they have to spread out in.

If we now apply this commonsense approach to the birthday problem, we might think that the chance of any of the hundred students sharing a birthday is equally slim, given that there are as many days to choose from as there are seats. Of course, there may well be some birthday buddies, but intuitively we would think that this is less likely than not.

Naturally, with a group of 366 people (leaving leap years aside), it needs no explaining how we can be certain that at least two will share a birthday. But things get interesting when we reduce the number of people.

In fact, incredible as it may seem, you need only fifty-seven people in the room for the probability of any two sharing a birthday to be as high as 99 per cent. That is, with only fifty-seven people, it's almost certain that two of them will share a birthday! This in itself sounds hard enough to believe. But as for the answer to the puzzle, the number above which it is 'more likely than not' that two share a birthday (that is, for the probability to be more than one-half) is considerably lower than fifty-seven. In fact, it is just *twenty-three* people!

Most people find this result very startling the first time they hear it, and continue to feel uneasy about it even when assured it's correct – because it is intuitively so difficult to believe. So let's go through the maths, which I will try to do as clearly as possible.

First, we keep the problem as simple as we can by assuming we are not dealing with a leap year, that all days in the year are equally probable for birthdays and that there are no twins in the room.

The mistake many people make is to think it is to do with comparing two numbers: the number of people in the room and the number of days in a year. Thus, since the twenty-three people have a choice of 365 days to have birthdays on, it seems far more likely than not that they will all avoid each other. But this way of looking at the problem is misleading. You see, for people to share birthdays we require *pairs* of people, not individuals, and we must consider the number of different pairs available. Let's start with the simplest case: with just three people there are three pairs: A–B, A–C and B–C. But with four people there are six pairs: A–B, A–C, A–D, B–C, B–D, C–D. With twenty-three people we find that there are 253 different pairs.* You see how much easier it becomes to believe that one of these 253 pairs of people will share a birthday from a choice of 365 days.

The way to work out the probability correctly is to start with one pair, keep adding people and see how the probability of birthday sharing changes. This is done by working out not the probability of

* There is a mathematical way of working this out called a binomial coefficient. In this case it is written as:
$$\binom{23}{2} = \frac{23 \times 22}{2} = 253$$

sharing, but rather the probability of each new person avoiding all other birthdays so far. Thus, the probability of the second person avoiding the birthday of the first is 364 ÷ 365, because he has all but one of the days in the year to pick from. The probability of the third person avoiding the birthdays of the first and second is then 363 ÷ 365. But we cannot forget about the first two people still having to avoid each other's birthdays too (the 364 ÷ 365 number). In probability theory, when we want to work out the chances of two different things happening at the same time, we must multiply the probability of the first and the probability of the second together. So the probability of the second person avoiding the birthday of the first, and of the third avoiding those of the first and second, is: 364/365 × 363/365 = 0.9918. Finally, if this is the probability of all three *avoiding* each other's birthdays, the probability of any two of the three *sharing* a birthday is 1 − 0.9918 = 0.0082. So the probability of sharing between just three people is pretty tiny, as you might expect.

We now carry on with this process – adding people one by one and building up the chain of multiplied fractions to work out the probability of everyone avoiding everyone else – until the answer we get drops below 0.5, i.e. 50 per cent. This is, of course, the point at which the probability of any pair *sharing* a birthday rises above 50 per cent. We find we need twenty-three fractions, hence twenty-three people:

$$\frac{364}{365} \times \frac{363}{365} \times \frac{362}{365} \times \frac{361}{365} \times \frac{360}{365} \times \ldots \ldots = 0.4927 \ldots$$

← 23 fractions multiplied together →

And so the probability of any two of the twenty-three people in the room sharing a birthday is:

$$1 − 0.4927 = 0.5073 = 50.73\%.$$

This puzzle has required some probability theory to solve it. The next one is, in a way, more straightforward. This I think makes it all the more incredible. It is my favourite veridical paradox because it is so easy to state, so easy to explain, and yet so hard to fathom.

The Monty Hall Paradox

This puzzle has its origins in Bertrand's Box Paradox and is an example of the power of what mathematicians call 'conditional probability'. It is based on an earlier puzzle called the Three Prisoners Problem, described by the American mathematician Martin Gardner in his 'Mathematical Games' column of the magazine *Scientific American* in 1959. But the Monty Hall Paradox is, I believe, a superior and much clearer adaptation. It is so called because it was first cast in the form of a scenario from the long-running US television game show *Let's Make a Deal*, presented by the charismatic Canadian, Monte Hall, who, on entering into show business, altered the spelling of his first name to Monty.

Steve Selvin is an American statistician and professor at the University of California in Berkeley. He is a renowned educator who has won awards for his teaching and mentoring. As an academic, he has applied his mathematical expertise to medicine, specifically in the field of biostatistics. However, he owes his worldwide fame not to these considerable achievements but to an amusing article he wrote on the Monty Hall Paradox. It was published in the February 1975 edition of an academic journal called *The American Statistician* and took up just half a single page.

Selvin could never have anticipated that his short article would have such a huge impact – after all, *The American Statistician* was a specialist journal read mainly by academics and educators – and indeed, fifteen years would pass before the problem he posed and solved burst into the popular consciousness. In September 1990 a reader of *Parade* magazine, a weekly publication boasting a US circulation in the tens of millions, submitted a puzzle to its 'Ask Marilyn' column, in which Marilyn vos Savant responds to readers' questions and solves their mathematical puzzles, brainteasers and logical conundrums. Vos Savant first rose to fame in the mid-1980s when she made it into *The Guinness Book of Records* for having the world's highest IQ (measured to be 185). The writer of this particular 'Ask Marilyn' entry was Craig F. Whitaker, and he essentially put to vos Savant a revised version of Selvin's Monty Hall Paradox. What followed was nothing short of remarkable.

The publication of the problem in *Parade* and Marilyn vos Savant's response brought it to nationwide, then worldwide, attention. Her answer, though completely counterintuitive, was, like Selvin's original solution, utterly correct. But it immediately spawned a host of letters to the magazine from incensed mathematicians eager to declare her wrong. Here are some extracts from three of them:

> As a professional mathematician, I'm very concerned with the general public's lack of mathematical skills. Please help by confessing your error and in the future being more careful.

> You blew it, and you blew it big! You seem to have difficulty grasping the basic principle at work here . . . There is enough mathematical illiteracy in this country, and we don't need the world's highest IQ propagating more. Shame!

> May I suggest that you obtain and refer to a standard textbook on probability before you try to answer a question of this type again?

> I am in shock that after being corrected by at least three mathematicians, you still do not see your mistake.

> Maybe women look at math problems differently than men.

Well, what a lot of angry people. And what a lot of subsequent egg on faces. Savant revisited the problem in a later issue and held her ground, arguing her case clearly and conclusively – as you might expect from someone with an IQ of 185. The story eventually made it on to the front page of the *New York Times* – and yet still the debate raged on (as you can see if you care to search for it online).

It may be starting to sound to you as though this paradox is so difficult to resolve that only a genius can really get their head round it. Not so. In fact, there are many simple ways of explaining it, and the internet is full of articles, blogs – even YouTube videos – that do so.

Anyway, enough of the teasing and historical rambling – let me get straight to the problem. I think it only fair to begin by quoting Steve Selvin's amusing original 1975 version in *The American Statistician*.

A PROBLEM IN PROBABILITY

It is 'Let's Make a Deal' – a famous TV show starring Monty Hall.

Monty Hall: One of the three boxes labelled A, B, and C contains the keys to that new 1975 Lincoln Continental. The other two are empty. If you choose the box containing the keys, you win the car.

Contestant: Gasp!

Monty Hall: Select one of these boxes.

Contestant: I'll take box B.

Monty Hall: Now box A and box C are on the table and here is box B (contestant grips box B tightly). It is possible the car keys are in that box! I'll give you $100 for the box.

Contestant: No, thank you.

Monty Hall: How about $200?

Contestant: No!

Audience: No!!

Monty Hall: Remember that the probability of your box containing the keys to the car is 1 in 3 and the probability of your box being empty is 2 in 3. I'll give you $500.

Audience: No!!

Contestant: No, I think I'll keep this box.

Monty Hall: I'll do you a favour and open one of the remaining boxes on the table (he opens box A). It's empty! (Audience: applause). Now either box C or your box B contains the car keys. Since there are two boxes left, the probability of your box containing the keys is now 1 in 2. I'll give you $1000 cash for your box.

WAIT!!!!

Is Monty right? The contestant knows that at least one of the boxes on the table is empty. He now knows it was box A. Does this knowledge change his probability of having the box containing the keys from 1 in 3 to 1 in 2? One of the boxes on the table has to be empty. Has Monty done the contestant a favour by showing him which of the two boxes was empty? Is the probability of winning the car 1 in 2 or 1 in 3?

Contestant: I'll trade you my box B for the box C on the table.

Monty Hall: That's weird!!

HINT: The contestant knows what he is doing!

Steve Selvin
School of Public Health
Univ. of California
Berkeley, CA 94720

In the above article, Selvin omits one crucial part of the problem (the relevance of which will become clear soon). He doesn't make clear that Monty Hall *knows* which box contains the keys – and so is always able to open an empty box. To be fair, he does state that Hall says: 'I'll do you a favour and open one of the remaining boxes on the table.' I take this to mean that Monty Hall knows perfectly well that the box he opens will be empty, but then I am familiar with the paradox. Although this might seem a trivial issue – after all, how could it possibly affect the odds as far as the contestant is concerned? – we shall see that the whole resolution rests on this point of what Monty Hall knows.

By the August 1975 edition of *The American Statistician* Selvin had to make this point clear because he found that, like Marilyn vos Savant fifteen years later, he came under criticism from other mathematicians who could not accept his solution. He wrote:

> I have received a number of letters commenting on my 'Letter to the Editor' in *The American Statistician* of February 1975, entitled 'A Problem in Probability.' Several correspondents claim my answer is incorrect. The basis to my solution is that Monty Hall knows which box contains the keys.

So that we can examine the problem more carefully, here is, with minor changes, the shorter and more famous version that appeared in *Parade*. In this version the three boxes are replaced with three doors:

> Suppose you're on a game show, and you're given the choice of three doors: A, B and C. Behind one door is a car, behind the

others, goats. You pick a door, say door A, and the host, who knows what's behind the doors, opens another door, say B, to reveal a goat. He says to you, 'Do you want to switch to door C?' Is it to your advantage to switch your choice of doors?

Of course, the assumption is that the contestant would much prefer to find a car than a goat. This isn't made clear and assumes the contestant is not a goat-loving cyclist.

Marilyn vos Savant's response, like Steve Selvin's years earlier, was that the contestant should always switch from his initial choice, because in doing so he *doubles* his chances of winning from 1 in 3 to 2 in 3. But how can this be? This is the crux of the Monty Hall Paradox.

Of course, most contestants faced with this choice will wonder if there is a trick here. Since the prize should be equally likely to be behind either door, why not go with their original hunch and stick with A? Surely, the car is now behind either door A or C with equal probability as far as the contestant is concerned, and it should make no difference whatsoever whether he sticks or switches.

This all sounds rather dubious and confusing, and you can see why even professional mathematicians were getting it wrong. Here then are several ways of explaining away the paradox.

Checking the probabilities

This is the most careful, methodical and watertight way of proving that indeed, by switching you double your chances of winning. Remember, you originally chose door A. Monty Hall, who knows where the car is, opens one of the other two doors to reveal a goat and offers you the opportunity to switch to C.

Consider first the case of sticking with door A.

The car could have been behind any one of the three doors with equal probability:

Figure 1.2. The Monty Hall Paradox: the problem

The prize is behind one of the three doors . . .

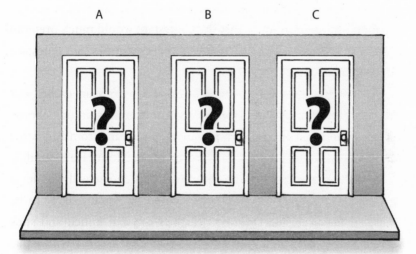

The game show host opens door B to reveal a goat. Do you now stick with your first choice of door A or do you switch to C?

Figure 1.3. The Monty Hall Paradox: the answer

If Monty Hall, who knows where the car is, opens door B to reveal a goat, then you have a 1 in 3 chance of winning the car if you stick to your original choice, A, compared with a 2 in 3 chance if you switch to C.

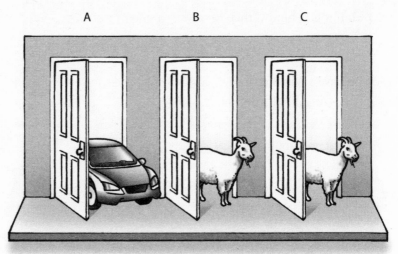

1 in 3 chance if you stick to A

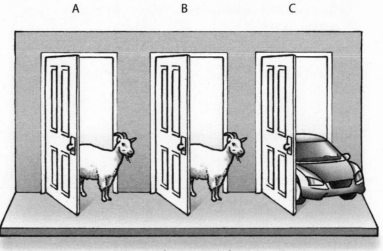

2 in 3 chance if you switch from A to C

❏ If it is behind A then it doesn't matter which of B or C is opened: you WIN.

❏ If it is behind B then door C is opened: by sticking with A: you LOSE.

❏ If it is behind C then door B is opened: by sticking with A: you LOSE.

So you have a 1 in 3 chance of winning the car if you stick.

Now consider the option of switching.

Again, the car could have been behind any one of the three doors with equal probability:

❏ If it is behind A then it doesn't matter which of B or C is opened: you LOSE.

❏ If it is behind B then door C is opened: by switching from A to B: you WIN.

❏ If it is behind C then door B is opened: by switching from A to C: you WIN.

So you have a 2 in 3 chance of winning the car if you switch.

Proof without maths – the commonsense approach

This is not really a proof in the strictest sense, but rather a non-mathematical way of making the solution easier to accept.

Consider if there were not three doors, but a thousand: one hiding the car and 999 hiding goats. You choose one of them at random – let us say it is door 777. Of course you may have chosen your particular door for any number of reasons, but the fact remains that, in the absence of psychic ability, you have a 1 in 1,000 chance of having picked the one hiding the car. What if Monty Hall, who knows where the car is, now opens every other door, in each case to reveal a goat, apart from door number 238. You can now see staring back at you 998 goats and two closed doors: your choice of door number 777 and the one remaining unopened door, number 238. Do you now stick or switch?

Would you not think that there's something fishy in that one particular door being spared by the host – some information he has that was unavailable to you when you made your own, random, choice? Remember, he knows where the car is. He sees you pick at random a door that is likely – the odds are overwhelming – to be hiding a goat. He then opens 998 other doors hiding goats. Would you not now feel compelled to switch to this final door? Of course you would, and you would be right: the car is almost certainly behind door 238, which has been deliberately spared by Monty.

To put this more mathematically, your initial choice splits the doors into two sets. Set 1 contains only the door you selected, and the probability that this hides the car is 1 in 3 (or 1 in 1,000 in the expanded case). Set 2 contains all remaining doors, and so the probability that the winning door is somewhere among them is 2 in 3 (or 999 in 1000). Opening one (or 998) of the doors in set 2, which is (are) known to hide a goat, thus having a probability of zero of hiding the car, leaves the set with just one unopened door, but the overall probability that this remaining door hides the car is still 2 in 3 (or 999 in 1000) because it has inherited the probability of the car being anywhere in that set. Opening the useless goat doors does nothing to change this probability that the car is behind a door in set 2.

The role of prior knowledge

You are no doubt convinced by now, but just in case you have any lingering doubts, here is another example that I feel clearly highlights this key distinction between having prior knowledge and not having it.

Let us suppose you wish to buy two kittens. You call the local pet shop and the owner informs you that two sibling kittens have just come in that day: a black one and a tabby one. You ask if they are boy or girl. Consider two different responses from the pet shop owner:

a) He tells you: 'I have only checked one of them and it's a boy.' With no further information, what is the probability that both kittens are boys?

b) He tells you: 'I have checked the tabby and it's a boy.' What now is the probability that they are both boys?

It turns out that the answers in the two cases are different. Although in both cases we know that at least one of the cats is a boy, it is only in the second case that we are told which one. That extra information is what changes the probability. Let's see how.

Begin with all four possible options:

	black	tabby
1	boy	boy
2	boy	girl
3	girl	boy
4	girl	girl

Consider now case (a): 'At least one of them is a boy.' This tells us it could be any one of the first three options: (1) both boys; (2) black is a boy and tabby is a girl; (3) black is a girl and tabby is a boy. So there is a 1 in 3 chance that they are both boys.

However, for case (b), when you are told specifically that the tabby is a boy, this extra information rules out option 2 in the table as well as option 4, leaving just two options: either both are boys, or the tabby is a boy while the black is a girl. Now the probability that both are boys is 1 in 2.

So you see the probability that both kittens are boys changes from 1 in 3 to 1 in 2 as soon as you know which of the two cats has been discovered to be a boy. This is precisely the same situation as that in the Monty Hall Paradox.

But wait a minute, I hear the hardened sceptic say: with the kittens story, the pet shop owner has passed on the extra information to you in order for you to work out the probabilities; Monty Hall did no such thing. This objection leads us to the final part of the explanation: at last, we can lay bare the issue that so confused readers of both Selvin's article in *The American Statistician* in 1975 and then Marilyn vos Savant's explanation in *Parade* magazine in 1990. We must, I'm afraid, return one last time to the Monty Hall Paradox.

Consider the situation in which Monty Hall does *not* know where the car is hidden. Now if he opens door B to reveal a goat then you are indeed left with equal probability that the car is behind door A or door C. How does this work? Well, imagine playing the game, with just the three doors, 150 times. Before each go, an independent judge moves the car randomly among the three doors such that even Monty Hall himself has no idea where it is. If you now pick a door and Monty Hall opens one of the other two at random, he will discover the car behind it, on average, a third of the time. Statistically, this corresponds to about 50 of the 150 attempts. In those 50 cases, of course, the game is over: you cannot carry on since you cannot win the car. This leaves the 100 times that Monty

Hall does reveal a goat behind door B. In each of these cases you are left with the 1 in 2 probability that the car is behind your original choice of door, and there is no reason to switch. That is, in 50 attempts the car will indeed be behind the door you picked and the other 50 times it will be behind C. Add to these the 50 cases in which the car was behind the door that Monty opened and we have three lots of 50, showing the equal likelihood that the car was behind each of the three doors.

Of course, had Monty known where the car was, he would not have needed to waste those 50 attempts by opening the door that hid it. So, to summarize, let us say you pick door A every time. In 50 of the 150 cases, the car is indeed behind A and you therefore have a 1 in 3 chance of winning it if you stick. Of the other 100 cases, half the time the car is behind C, so Monty opens door B, and the other half of the time it is behind B, so he opens C. In all 100 of these cases, he opens the door hiding the goat, leaving the car invisible behind the other one. So by switching in each of the 150 cases you will win the car on 100 occasions – 2 in 3 of the total.

Suck it and see

In her final column on the problem, Marilyn vos Savant announced the results of more than a thousand school experiments that had taken place to test the problem. Almost every one of the results concluded that switching was the right option. This 'suck it and see' way of resolving the paradox is also the method I had to resort to some years ago when trying to explain it to a friend. On a long car journey travelling to a filming location for a television science documentary I was making for the BBC, I recounted the paradox to my cameraman, Andy Jackson. At that point I had not, I must confess, honed the arguments and explanations I've set out here, and so resorted to pulling out a deck of cards to demonstrate. I picked three cards, one red and two black, shuffled them, then placed them face down in a row on the car seat between us. I then carefully took a peek under each to see where the red one was. I asked Andy to pick the red card without turning it over. I then turned over one of the other two that I knew was black and gave him the option of sticking or switching. It took no more than about twenty attempts to show him that it was roughly twice as likely that he would pick the red one if he switched. He didn't quite get how, but was at least convinced I was right.

I hope Andy reads this chapter and finally understands – as I hope you do – why.

And so, enough of this frivolity – nine proper physics problems await us.

TWO

Achilles and the Tortoise

'All motion is an illusion'

THE FIRST OF our nine paradoxes goes back two and a half millennia and, given how long we've had to mull it over, you will not be surprised to hear that it has been thoroughly understood and explained. And yet most people encountering it for the first time still scratch their heads in bafflement. It is known as the Paradox of Achilles (or the Problem of Achilles and the Tortoise) and is just one of a series of problems set by the Greek philosopher Zeno in the fifth century BC. As an example in pure logic it couldn't be simpler. But don't be fooled; in this chapter we will consider several of Zeno's paradoxes and finish off by bringing his ideas right up to date with a version of one that can only be explained using quantum theory. Well, I never said I would go easy on you.

But let us begin with the most famous of Zeno's paradoxes. A tortoise is given a head start in a race against the swift-running Achilles such that it reaches some point (let's call it point A) along the route by the time Achilles sets off. Since Achilles runs much faster than the tortoise walks, he will very soon get to point A. However, by the time he gets there the tortoise has already moved on a short distance to a further point, which we will call point B. By the time

Achilles reaches point B, the tortoise has moved on to point C, and so on. So while Achilles is clearly catching up with the tortoise, and the gap between them becomes a bit smaller at each stage, it seems he will never actually overtake it. Where are we going wrong?

When it comes to being clever, mastering logical conundrums and brainteasers, or just deep thinking in general, you can't beat the ancient Greeks. In fact, so sharp were these philosophers of antiquity, so insightful in their logic, we tend to forget that they lived over two thousand years ago. Even today, when we wish to give examples of genius, along with the ever-popular Einstein we often invoke such familiar names as Socrates, Plato and Aristotle as representing the ultimate in intellectual brilliance.

Zeno was born in Elea, an ancient Greek town in what is now south-western Italy. Little is known about his life and work other than that he was a student of another Elean philosopher, Parmenides. Together with a third man from the same town, Melissus, they formed what is now called the Eleatic movement. Their philosophy was that you should not always trust your senses and sense experience in order to understand the world, but should rely ultimately on logic and mathematics. On the whole, this is a sensible approach; but, as we shall see shortly, it would lead Zeno down the wrong path.

From what little we know of his ideas it would seem that Zeno did not have many positive views of his own, but was instead committed to demolishing the arguments of others. Despite this, Aristotle himself, who lived a century after Zeno, regarded him as the founder of a type of debate called 'dialectic'. This is a form of civilized discussion at which the ancient Greeks – particularly men like Plato and Aristotle – excelled, using logic and reason to resolve disagreements.

Only one short piece of Zeno's original work survives today, so what we know about him derives from the writings of others, notably

Plato and Aristotle. At around the age of forty Zeno travelled to Athens, where he met the young Socrates. Later in life he became active in Greek politics, and was eventually imprisoned and tortured to death for his part in a conspiracy against the local ruler in Elea. One story about him is that he bit off his own tongue and spat it out at his captors rather than betray his co-conspirators. But he is most famous for a series of paradoxes that are brought to us by Aristotle in his great text the *Physics*. There are believed to have been around forty in total, but only a handful survive.

All Zeno's paradoxes – the four most famous of which are known by the names Aristotle gave them: the Achilles, the Dichotomy, the Stadium and the Arrow – are centred on the idea that nothing ever changes; that motion is just an illusion and that time itself does not really exist. Of course, if there was one thing the Greeks excelled at it was philosophizing, and grand statements like 'all motion is an illusion' are just the sort of provocative abstraction they were famous for. Today we can demolish these paradoxes scientifically; but they are such fun that it is well worth revisiting them here. In this chapter I will consider them in turn and show how each one can be resolved with a little careful scientific analysis. Let's begin with the one I've already outlined.

Achilles and the Tortoise

This is my favourite of Zeno's paradoxes because at first glance it seems so perfectly logical, and yet it defies logic in an unexpected way. Achilles is the greatest warrior in Greek mythology, endowed with tremendous strength, bravery and military skill. Part human, part supernatural being – his parents were King Peleus of Thessaly and a sea nymph named Thetis – he figures prominently in Homer's *Iliad*, which tells the story of the Trojan War. It was said that even as a young boy he could run fast enough to catch a deer and was strong enough to kill a lion. So Zeno was clearly contrasting extremes when he chose this mythical hero to race against the ponderous tortoise.

The paradox is based on the even older fable of the hare and the tortoise, attributed to another ancient Greek, by the name of Aesop, who lived about a century before Zeno. In the original fable, the tortoise is ridiculed by the hare and so challenges him to a race, which the tortoise duly wins thanks to the hare's arrogance in thinking he can afford to stop for a nap halfway through, only to awaken too late to catch the tortoise.

In Zeno's version, the fleet-footed Achilles takes on the role of the hare. Unlike the hare, he is completely focused on the task; but he does give the tortoise a head start, and this seems to be his undoing, as the tortoise appears always to win the race, however long it is, albeit probably by the ancient Greek equivalent of a photo finish. According to Zeno's account, however fast the hero runs, and however slowly his opponent plods, Achilles will never overtake the tortoise. Surely this cannot be what happens in reality?

This was a serious conundrum to the Greek mathematicians, who had no real concept of what we call a converging infinite series, or indeed of the notion of infinity itself (ideas I will explain briefly to you in a moment). Aristotle, who was certainly no slouch when it came to thinking about such matters, regarded Zeno's paradoxes as 'fallacies'. The problem was that neither Aristotle nor anyone else in ancient Greece properly understood one of the most basic algebraic formulas in physics: speed equals distance divided by time. Today we can do much better.

The statement 'will never overtake the tortoise' is of course wrong, since the ever-decreasing distances that are being considered with each stage (between points A and B, and between B and C, and so on), also involve ever-decreasing time intervals, and so even an infinite number of stages does not imply an infinite length of time. In fact, the stages all add up to a finite time: the time it takes for Achilles to reach the tortoise! The confusing thing about the paradox is that most people don't appreciate that adding up an infinite sequence of numbers does not necessarily lead to an infinite result. Strange though it might sound, an infinite number of stages can be completed in a finite time, and the tortoise will be reached and overtaken easily enough, as logic insists. The solution relies on what mathematicians call a geometric series.

Consider the following example:

$$1 + \tfrac{1}{2} + \tfrac{1}{4} + \tfrac{1}{8} + \tfrac{1}{16} + \tfrac{1}{32} \ldots$$

It is clear that you could keep adding smaller and smaller fractions for ever, with the total getting closer and closer to the value 2. Try this by drawing a line and then dividing it in half, then taking the right-hand side and chopping that in half, and so on until the fractions get so tiny that you can no longer mark them separately on the paper. If half of

Figure 2.1. A converging infinite series
Summing an infinite number of ever-decreasing lengths – adding lengths for ever – doesn't mean the final answer is infinity since the lengths are getting shorter all the time.

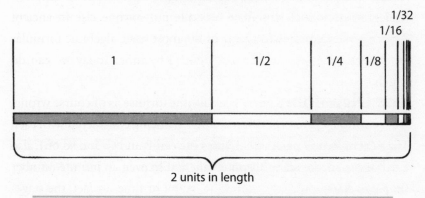

2 units in length

the line is one unit in length (it doesn't matter if this is a centimetre, an inch, a metre or a mile), then by adding successive fractions, as in the series written above, we converge on the total length of two units.

A good way to apply this to the paradox is to consider not the points at which Achilles and the tortoise respectively have arrived at any stage, but rather the ever-decreasing gap between them. Since each is moving at a constant speed, this gap is decreasing at a constant rate too. For example, if Achilles gives the tortoise a 100-metre start and then proceeds to catch it up at a rate of 10 metres per second, how does this pan out according to Zeno? Well, the distance between them is halved after five seconds. The remaining distance is itself halved after two and a half seconds, then what is left is halved after one and a quarter seconds and so on. We could, if we wished, keep adding the ever-smaller distances covered during these ever-decreasing time intervals, but the fact remains that if Achilles is catching the tortoise up at a rate of 10 metres per second then he will overtake it in 10 seconds, which is the time he needs to close that initial gap of 100 metres to zero. And this value of 10 seconds is just the number we would get if we added up 5 seconds + 2.5 seconds + 1.25 seconds + 0.625 seconds . . . and so on until the next number to add is so small we are happy to call it a day (at 9.9999 . . . seconds). After 10 seconds Achilles will then of course streak on ahead of the tortoise as expected (unless he decides to stop for a beer along the way – not something Zeno felt was necessary to clarify in his argument).

The Dichotomy

The next of Zeno's paradoxes refutes the reality of motion itself and is a variation on the same theme as the Paradox of Achilles. It is very simple to state:

In order to reach your destination you must first cover half the distance, but in order to cover half the distance you must first cover a quarter of the distance, and in order to cover a quarter of the distance you must cover an eighth of the distance, and so on. If you can keep chopping the distances in half for ever, then you never reach that very first distance marker, and so you never actually start your journey. What's more, this never-ending sequence of ever-shorter distances is infinite. So to complete the journey requires you to complete an infinite number of tasks. So you could never finish it. If you cannot start the journey and would never finish it anyway, then motion itself is impossible.

We learn about this paradox from Aristotle, who knows that it is nonsense but searches for the logical argument with which he can refute it conclusively. After all, it is quite obvious that there is such a thing as motion. However, Zeno was applying a form of argumentation called *reductio ad absurdum*, whereby an idea is pushed and pushed until it leads to a logically absurd conclusion. We must also remember that Zeno was no mathematician. He was arguing with reference only to pure logic, and that is often just not enough. Other Greek philosophers resorted to a more direct and pragmatic approach in refuting Zeno's arguments about the illusion of motion. One of them was Diogenes the Cynic.

Our word 'cynicism' has its origin in an idealistic philosophical movement of ancient Greece. The Greek Cynics seem to have been a nicer bunch than the modern connotations of their name suggest: they rejected wealth, power, fame, even possessions, and chose instead to lead a simple life free of all the traditional human vices. They believed that all humans were equal and that the world belonged in equal measure to everyone. Probably the most famous of the Cynics was Diogenes, who lived during the time of Plato in the fourth century BC. This philosopher is responsible for some wonderful quotes, such as 'Blushing is the colour of virtue,' 'Dogs and philosophers do the greatest good and get the fewest rewards,' 'He who is most content with the least has the most' and 'I know nothing, except the fact of my ignorance.'

Diogenes took the teachings of Cynicism to their logical extremes. He seems to have made a virtue of poverty and spent years living in a tub in an Athens marketplace. He became famous for being, well, cynical about everything, particularly much of the philosophical teaching of the time, even that of eminences such as Socrates and Plato. So you can imagine what he thought of Zeno and his paradoxes. On hearing about Zeno's Dichotomy Paradox regarding the illusion of motion, he simply stood up and walked off, thus demonstrating the absurdity of Zeno's conclusions.

While we may applaud Diogenes for his practical approach, we still need to investigate a little more carefully where Zeno's logic is breaking down. And that turns out not to be so difficult – after all, we've had over two thousand years to figure it out. In any case, while you may feel that sheer common sense is sufficient to dismiss Zeno's paradox, I do not. I have spent most of my life working and, more importantly, thinking as a physicist, and I am not satisfied with mere commonsensical, philosophical or logical arguments that refute the Dichotomy. I need watertight physics – which, for me, does a far more convincing job.

What we need to do is to convert Zeno's argument about distance into one about time. Assume you are already moving at a constant speed at the moment in time when you are at the starting point of the journey to be covered. The notion of speed, which Zeno would not have understood very well, means covering a certain distance in a finite time. The shorter the distance you must cover, the shorter the time interval needed to cover it, but whenever you divide the first number by the second they always give the same answer: your speed. By considering shorter and shorter distances that must be covered for you to have begun your journey, you are also considering shorter and shorter intervals of time. But time marches on regardless of how we might wish to split it up artificially into these ever-decreasing periods. Thinking about time, rather than space, as a static line that can be subdivided indefinitely is fine (and we often think of time in this way when solving problems in physics), but the crucial point is that the way we *perceive* time is not as a static line in the same way as we can view lines in space. We cannot take ourselves outside of time's stream. Time marches on regardless – and so we move.

If we consider the situation from the point of view of someone not already moving, but starting from rest, there is just one more bit of physics we need to think about. This is something we all learn about at school (and most of us, no doubt, promptly forget). It is referred to as Newton's second law, which states that to make an object begin to move, a force needs to be applied to it. This will cause it to accelerate – to alter its state from being at rest to being in motion. But once it is moving, the same argument applies: namely that, as time goes by, the distances covered are based on the moving object's speed, which need not be constant. The Dichotomy argument is then an abstract irrelevance that has nothing to say about true motion in the physical world.

I should make one final remark before moving on. Albert Einstein's theory of relativity teaches us that maybe we should not dismiss the Dichotomy Paradox so confidently. According to Einstein, time *can* be regarded in a similar way to space – indeed, he refers to time as the fourth axis, or fourth dimension, of what is called space–time. This suggests that maybe the flow of time is just an illusion after all – and, if it is, then so is motion. But I would argue that, despite the success of relativity theory, this conclusion takes us away from physics and into the murky waters of metaphysics – abstract ideas that don't have the solid backing of empirical science.

I am not suggesting that Einstein's theory of relativity is wrong; of course not. It is just that Einstein's ideas only really manifest themselves when things start to move very fast – close to the speed of light. At normal everyday speeds we are quite within our rights to ignore such 'relativistic' effects and think of time and space in the familiar commonsense way we are used to. After all, if we push Zeno's argument to its logical extreme, then it is in fact wrong to say that time and space are infinitely divisible into ever-smaller but still discrete intervals and distances. At some point things get so small that quantum physics comes into effect, when time and space themselves become fuzzy and indefinable and it no longer makes sense to chop them up into smaller pieces. Indeed, motion itself is a little illusory in the quantum domain of atoms and subatomic particles. But that is not what Zeno had in mind.

While fun to explore and discuss in this context, neither quantum physics nor relativity theory is needed to explain away Zeno's Dichotomy. Using such ideas from modern physics to argue that all motion is an illusion is to miss the point and brings us dangerously close to turning physics into mysticism. So let's not make things more complicated than they need to be. There will be plenty of time for such craziness later on in the book, believe me.

The Stadium

And so we move swiftly on. A related paradox of Zeno's that plays with the notion of speed is known as the Moving Rows Paradox. It is somewhat obscure and reaches us via Aristotle, who called it the Stadium Paradox. I will try to describe it as succinctly as possible.

Figure 2.2. The Moving Rows Paradox

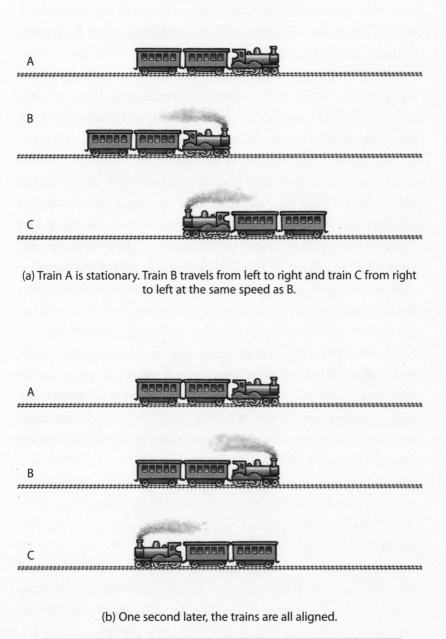

(a) Train A is stationary. Train B travels from left to right and train C from right to left at the same speed as B.

(b) One second later, the trains are all aligned.

Imagine there are three trains, each consisting of one engine and two carriages. The first train is standing still at a station. The second and third trains don't stop at the station but are moving at an equal and constant speed in opposite directions to each other, B from the west and C from the east.

At a given moment in time the trains are positioned as in Figure 2.2(a). Then, one second later, they are aligned as in Figure 2.2(b). The problem, according to Zeno, concerns the motion of train B: in one second, it has moved the length of one carriage past train A, but at the same time it has moved the length of two carriages past train C. The paradox is that train B has moved a distance and twice that distance at the same time. Zeno seems to have been aware that these are just relative distances, and so attempted to cast the paradox in terms of time instead. Dividing each of the two distances by the constant speed of train B, we arrive at two different time periods, one double the duration of the other. But both paradoxically seem to describe how long it takes to get from the situation in the upper diagram to the aligned one in the lower diagram!

Resolving this apparent paradox is easy, since we can see what is going wrong in the reasoning. There is such a thing as relative speed, of course, so that we cannot say that B is travelling at the same speed relative to the moving C as it is relative to the stationary A. Did Zeno know this, and was he simply making a more subtle point about the illusory nature of motion? It is not clear; but, as every schoolboy and schoolgirl should be able to appreciate, there is really no paradox here at all. B passes C at twice the relative speed that it passes A, and so of course it will pass two carriages of C in the same time it takes to pass one carriage of A.

The Arrow

Like the Dichotomy, this is another paradox centred on the idea that true motion is just an illusion. It is stated thus by Aristotle: 'If everything when it occupies an equal space is at rest, and if that which is in locomotion is always occupying such a space at any moment, the flying arrow is therefore motionless.'

Eh? Well, let me try to put it more clearly.

An arrow in flight has, at any given instant in time, a certain, fixed position – as we would see in a snapshot. But if we only see it at this instant, it will be indistinguishable from a truly motionless arrow in the same position. So how can we ever say an arrow is in motion? Indeed, since time is made up of a sequence of consecutive instants, in each of which the arrow is motionless, it never moves.

The paradox, of course, is that we know there is such a thing as motion. Of course the arrow moves. So where is the logical error in Zeno's statement?

Time can be considered to be made up of a sequence of infinitesimally short 'moments', which we can think of as the smallest possible, indivisible, intervals of time. As a physicist, I can see the problem with Zeno's argument. If these indivisible instants are not exactly of zero duration (true snapshots) then the arrow will be in a slightly different position at the start of each one from its position at the end, and it cannot therefore be said to be at rest. On the other hand, if such instants are truly of zero duration, then it doesn't matter how many of them sit together side by side, they will never add up to a non-zero interval of time: we can add zero to itself as many times as we like and the answer is still zero. So Zeno's argument that a finite duration of time is just made up of a sequence of such consecutive instants is wrong.

It would take advances in mathematics as well as physics for this paradox to be finally laid to rest. More specifically, it was an understanding of calculus, the field of mathematics developed by Isaac Newton and others in the seventeenth century which describes how to add up tiny quantities in order to describe the notion of change correctly, that would finally clarify Zeno's naïve ideas.

But there is a sting in the tail. In 1977, two physicists at the University of Texas published a surprising research paper that suggested Zeno's Arrow Paradox might have been laid to rest prematurely. Their names were Baidyanaith Misra and George Sudarshan, and the title of their paper was 'The Zeno's Paradox in Quantum Theory'. Physicists around the world were intrigued. Some thought the work silly, while others rushed to try to test the idea. But before I go any further, let me say as little as I feel I can get away with at this early stage of the book about the weird and wonderful set of ideas that underpins quantum mechanics.

Zeno's Paradox and quantum mechanics

Quantum mechanics is the theory that describes the workings of the microscopic world – by which I do not mean the tiny world only visible under a microscope, but rather the far, far tinier world of atoms and molecules and the subatomic particles (the electrons, protons and neutrons) of which they are made up. Indeed, quantum mechanics is the most powerful, important and fundamental mathematical set of ideas in the whole of science. It is remarkable for two seemingly contradictory reasons (almost a paradox in itself!). On the one hand, it is so fundamental to our understanding of the workings of our world that it lies at the very heart of most of the technological advances made in the past half-century. On the other hand, no one seems to know exactly what it means.

I must make it clear from the outset that the mathematical theory of quantum mechanics is not in itself weird or illogical; on the contrary, it is a beautifully accurate and logical construction that describes nature superbly well. Without it we would not be able to understand the basics of modern chemistry, or electronics, or materials science; we would not have invented the silicon chip or the laser; there would be no television sets, computers, microwaves, CD and DVD players or mobile phones, let alone much else that we take for granted in our technological age.

Quantum mechanics accurately predicts and explains the behaviour of the very building blocks of matter with extraordinary accuracy. It has led us to a very precise and almost complete understanding of how the subatomic world behaves, and how the myriad particles interact with each other and connect up to form the world we see around us, and of which we are of course a part. After all, we are ultimately just a collection of trillions of atoms obeying the rules of quantum mechanics and organized in a highly complex way.

These strange mathematical rules were discovered in the 1920s. They turn out to be very different from the rules that govern the more mundane everyday world we are familiar with: the world of objects we see around us. Near the end of the book I will explore just how strange some of these rules are when we consider the Paradox of Schrödinger's Cat. For now, I wish to focus on one particularly strange feature of the quantum world, namely that an atom will behave differently if left to its own devices from how it behaves when it is being 'observed' – by which I mean when it is being monitored in some way: poked or prodded, knocked or zapped. This feature of the quantum world is still not fully understood, partly because it is only now becoming clear what exactly constitutes 'observation' in this sense. This is known as the 'measurement problem' and is still an active area of scientific research today.

The quantum world is ruled by chance and probability. It is a place where nothing is as it seems. If left alone, a radioactive atom will emit a particle, but we are unable to predict when this will take place. All we can ever do is work out a number called the half-life. This is the time it takes for half of a large number of identical atoms to 'decay' radioactively. The larger the number, the more accurate we can be about this half-life, but we can never predict in advance which atom in the sample will go next. It is very much like the statistics of tossing a coin. We know that if we toss a coin again and again, then half the time it will end up heads and the other half tails. The more times we toss it, the more accurate this statistical prediction will be. But we can never predict whether the very next toss of the coin will be heads or tails.

The quantum world is probabilistic in nature not because quantum mechanics as a theory is incomplete or approximate, but rather because the atom itself does not 'know' when this random event will take place. This is an example of what is called 'indeterminism', or unpredictability.

Misra and Sudarshan's paper, which was published in the *Journal of Mathematical Physics*, describes the astonishing situation whereby a radioactive atom, if observed closely and continuously, will never decay! The idea can be summed up perfectly by the adage 'the watched pot never boils', which was first used, as far as I can tell, by the Victorian writer Elizabeth Gaskell in her 1848 novel *Mary Barton* – although it is the sort of saying that probably dates back much further. The notion has its origins, of course, in Zeno's Arrow Paradox and our inability to detect motion by considering a snapshot of a moving object in an instant of time.

But how and why might this happen in reality? Clearly the saying about the watched pot is nothing more than a simple lesson in patience: you cannot make a kettle boil any more quickly by staring at it. However, Misra and Sudarshan seemed to be suggesting that when it comes to atoms you really do influence how they behave by watching them. What's more, this interference is unavoidable – the act of looking will inevitably alter the state of the thing you are looking at.

Their idea goes to the very heart of how quantum mechanics describes the microscopic world: as a fuzzy, ghostly reality in which all sorts of weird goings-on seem to take place routinely when it's left alone – an idea we will revisit in Chapter 9 – none of which we are ever able to detect actually happening. So an atom that would, if left to its own devices, spontaneously emit a particle at any moment will somehow remain too shy to do so if it is being spied upon, so that we can never actually catch it in the act. It is as though the atom has been endowed with some kind of awareness, which is crazy. But then the quantum world is crazy. One of the founding fathers of quantum theory was the Danish physicist Niels Bohr, who in 1920 set up a research institute in Copenhagen to which he attracted the greatest scientific geniuses of the time – men such as Werner Heisenberg, Wolfgang Pauli and Erwin Schrödinger – to try to unlock the secrets of the tiniest building blocks of nature. One of Bohr's most famous sayings was that 'if you are not astonished by the conclusions of quantum mechanics then you have not understood it'.

Misra and Sudarshan's paper was entitled 'The Zeno's Paradox in Quantum Theory' because of its origins in the Arrow Paradox. However, it is fair to say now that, while its conclusion remains some-what controversial, it is, for most quantum physicists, no longer a paradox. In the literature today it is referred to more commonly as the 'Quantum Zeno Effect', and has been found to apply far more widely than in the situation described by Misra and Sudarshan. A quantum physicist will happily tell you that the effect can be explained by 'the constant collapse of the wave function into the initial undecayed state', which is the sort of incomprehensible geeky gobbledygook one should expect from such people – I should know, I am one of them. But I don't think I will pursue this line of thought in any more detail here, just in case you are nervously wondering what you've let yourself in for.

This recent discovery that the Quantum Zeno Effect is pretty much ubiquitous is down to a better understanding among quantum physicists of how an atom responds to its surroundings. A big breakthrough was made when scientists at one of the world's most prestigious laboratories, the National Institute of Standards and Technology in Colorado, confirmed the Quantum Zeno Effect in a famous experiment in 1990. The experiment took place within the wonderfully named Division of Time and Frequency, which is best known for setting the standards for the most accurate measurement of time. Indeed, scientists there have recently built the world's most accurate atomic clock, precise to within one second every three and a half *billion* years – close to the age of the Earth itself.

One of the physicists working on these incredibly high-precision clocks is Wayne Itano. It was his group who designed the experiment to test whether the Quantum Zeno Effect could really be detected. This involved trapping several thousand atoms in a magnetic field and then zapping them delicately with lasers, forcing them to give up their secrets. Sure enough, the researchers found clear evidence of the Quantum Zeno Effect: under constant watching, the atoms behaved very differently from what the scientists had expected.

One final twist: there is now evidence for the opposite effect, something called the 'Anti-Zeno Effect', which is the quantum equivalent of staring at a kettle and making it come to the boil more quickly. While still somewhat speculative, such research goes to the heart of some of the most profound and possibly important areas of science in the twenty-first century, such as working towards building what is called a quantum computer. This is a device that makes direct use of some of the strange behaviour of the quantum world in order to carry out its calculations far more efficiently.

I am not sure what Zeno of Elea would have made of this revival of his paradoxes, or of his name being attached to a remarkable phenomenon in physics nearly two and a half thousand years later. Here, the paradox has nothing to do with tricks of logic and everything to do with the even stranger tricks that nature seems able to play down at the tiny scale of atoms – tricks that we are only beginning to understand.

Zeno's paradoxes have taken us from the very birth of physics to cutting-edge ideas in the twenty-first century. All the other paradoxes in this book arose somewhere in between. In resolving them we will have to travel to the furthest reaches of the Universe and explore the essence of space and time themselves. Hold on tight.

THREE

Olbers' Paradox

Why does it get dark at night?

SEVERAL YEARS AGO I was on holiday with my family and a group of
friends in France. We were staying in an idyllic country farmhouse
in the Limousin region of the Massif Central, one of the most
sparsely populated areas of the country. Late one evening, with the
children asleep, we adults were sitting outside enjoying a last glass
or three of the local red wine, looking up at the clear, sparkling
night sky and commenting on how France is big enough for one
still to find uninhabited swathes of countryside where the light
pollution is low, and how unaccustomed we were, living in the
densely populated south-east of England, to being able to see
so many stars overhead. Most impressive of all was a large streak
of faint, diffuse light smeared across the sky, looking like a wispy
cloud.

But a cloud would have obscured the distant stars in our line of
sight from view – and we could clearly see just as many stars over-
laying this wispy streak as elsewhere in the sky. It seemed as though it
had to be something *behind* the distant stars. Being the resident scientist in the group, I was eager to point out that what we were seeing was
the central disc of our Milky Way galaxy edge on, and that this patch

of light was vastly further away than all the individual stars we could see. Several of my friends admitted, to my astonishment, that they had never seen it before, and were fascinated to hear me explain that it consisted of the billions of stars that make up the bulk of our galaxy and are too far away and too faint for us to see as individual pinpricks of light.

Of course, not every one of the pinpricks of light we do see in the night sky is a star. The brightest objects of all (discounting the Moon) are our planetary neighbours Venus, Jupiter and Mars. These glow because they reflect the light of the Sun, which is hidden from our view on the other side of the Earth during the night. The nearest stars to us beyond our solar system are several light-years away. Remember that, confusingly, a light-year is a unit of distance, not time. It is the distance light travels in one year, and comes in at just under 10 trillion kilometres. To put this into a context that I hope is easier to grasp, the 150 million kilometres that separate us from the Sun amount to just 0.000016 of a light-year. In fact, it is more sensible to say that the distance between the Earth and the Sun is 8.3 light-minutes, since that is how long (just over eight minutes) it takes light to cover the distance.

The next nearest star to us, after the Sun, is Proxima Centauri at just over four light-years away. But it is not the brightest star in the sky; that title goes to Sirius, which is twice as far away. Only the Moon, Jupiter and Venus are constantly brighter than Sirius, which can be seen wherever you are on the planet, unless you live several hundred miles north of the Arctic Circle. It forms, along with Betelgeuse and Procyon, one of the three vertices of the Winter Triangle of stars as viewed from the Northern Hemisphere. To find it, locate the three stars that make up the belt of Orion and follow them downwards. It's hard to miss.

Other bright stars include the very distant but massive Rigel, a blue supergiant, seventy-eight times as big as the Sun and 85,000 times as bright, making it the most luminous star in our local region of the galaxy. We don't see it shine as brightly as other stars in the sky, like Sirius, because it is so far away (between 700 and 900 light-years from the Earth). At about the same distance away, but even bigger than Rigel, is the slightly less luminous red supergiant Betelgeuse. This great star is 13,000 times as bright as the Sun and a thousand times as big – so huge that, if it were to replace the Sun at the centre of our solar system, it would engulf the orbits of Mercury, Venus, the Earth, Mars and Jupiter!

Once astronomers began to use telescopes, enabling them to see further out into the heavens than they ever could with the naked eye, they realized that stars are not distributed evenly throughout the Universe, but are instead clustered together in galaxies, each of which is like an immense star city; and that these galaxies are separated by unimaginably vast empty expanses of space. All the stars we can see in the sky (including Sirius, Rigel and Betelgeuse) are part of our own galaxy, the Milky Way. Indeed, they are all in our local tiny neighbourhood of the Milky Way.

Under perfect conditions (right part of the world, right time of year) we should be able to make out several thousand stars with the naked eye, and hundreds of thousands with a decent small telescope. Yet even this figure is a tiny fraction (less than 1 per cent) of all the stars in the Milky Way, which contains between 200 and 400 billion stars – that's around fifty stars for every human alive in the world today.

This is why the disc of the Milky Way appears as a continuous patch of faint light across the night sky. The galactic core is about 25,000 light-years from the Earth, while the galaxy as a whole has a diameter of 100,000 light-years. Such distances mean that individual stars are far too faint to resolve into clearly defined points of light, so that all we can see is the cumulative light from billions of them together.

The stars are not uniformly spread throughout the galaxy. Unlike our Sun, which sits alone, most stars come in pairs or groups orbiting each other. Some young stars hang together in their hundreds in loose open clumps, while larger groups, thousands strong, can be found in what are known as globular clusters.

We certainly cannot distinguish individual stars in other galaxies. In fact, it is almost impossible to see other galaxies at all without the help of powerful telescopes. Even our nearest neighbour galaxies, Andromeda and the Magellanic Clouds, are barely visible to the naked eye, appearing only as diffuse patches of very faint light.

The Andromeda galaxy, which is a little bigger than ours, is 2 million light-years away. If we were to shrink the Milky Way down to the size of the Earth, then Andromeda would be as far away as the Moon. Andromeda contains about 500 million (or half a billion) stars. I recall the thrill of first seeing it through a telescope, appearing as a faint fuzzy spiral. What hit me was the fact that I was seeing Andromeda not as it actually is now, but as it existed 2 million years ago. It was the light that had left it then, long before humans existed on Earth, that was only now entering my eye, thus completing its long journey. I felt weirdly privileged to have been there at that very moment to capture those photons of light as they made contact with my retina, triggering electrical signals that were sent to the neurons in my brain, making me conscious of what I was seeing.

Physicists often tend to think in this strange way.

Not only do stars form clusters within galaxies, galaxies themselves also group together in clusters. Our galaxy is one of about forty that together are known as the Local Group and also include the Large and Small Magellanic Clouds and Andromeda. Astronomical measurements have now reached such a degree of precision and sophistication, with ever more powerful telescopes being built, allowing us to probe ever deeper into space, that we now know that galaxy clusters are themselves grouped together into what are known as superclusters. Our Local Group is in fact part of the Local Supercluster. How far out do the contents of our universe extend? Is it, in fact, infinite? We simply do not know. But the question has plagued astronomers for centuries and leads to our next paradox.

As we stare out into the night sky, we may ask a very profound question:

Why does it get dark at night?

You might think that this is a rather trivial question. After all, even a child knows that night falls when the Sun 'sets' below the horizon, and that since there is nothing else in the night sky anywhere near as bright as the Sun we have to make do with the feeble reflected light from the Moon and the even more feeble light emanating from the distant planets and stars.

And yet it turns out this question is far more significant than it first appears. Indeed, astronomers puzzled over it for hundreds of years before they found the correct answer. It is known today as Olbers' Paradox.

Here, then, is the problem. We have good reason to believe that even if the Universe is not infinite in size (and it may well be), it is so enormous that to all intents and purposes it goes on for ever. Thus, in every direction we care to look out into space, we should see a star, and the sky should be even brighter than it normally becomes during the day; in fact, it should be so bright, all the time, that it should not even matter whether it is day or night according to our Sun.

Let's look at it another way. Imagine that you are standing in the middle of a very large forest – so large, in fact, that you can assume it extends to infinity in all directions. Now try shooting an arrow horizontally in any direction. In this idealized situation the arrow must be allowed to keep on going in a straight line without ever dipping down to the ground until it hits a tree trunk. Even if it misses all the closer trees, the arrow must eventually hit one. Since the forest is infinite, there will always be a tree in the flight path of the arrow, however far away that tree may be.

Now imagine that our universe goes on for ever with an infinity of stars evenly distributed throughout it. The light that reaches us from these stars is like the example of the arrow, but in reverse. For no matter where we look in the sky, we should always see a star in our line of sight. So there would not be any gaps where we do not see a star, and the whole sky should be as bright as the surface of the Sun, all the time.

When you first consider this dilemma, you might raise two points, both of which featured in my introduction to this paradox. First, you might ask: wouldn't very distant stars simply be too faint for us to see? And second: stars are *not* evenly spread out in the sky, are they? Aren't they bunched up in clusters, and those clusters grouped into galaxies? It turns out neither of these issues matters. In response to the first point, while it is true that more distant stars appear dimmer than closer ones, the corresponding patch of sky they are in would encompass a much larger volume since it is further away, and so would contain more stars. A relatively simple bit of geometry that I shall discuss a little later on in this chapter shows that these two effects balance out exactly: for any given patch of sky, the fewer, nearer stars would have a total brightness equal to that of the more numerous, more distant stars. On the second point, it is indeed true that stars are not spread out evenly but are collected together in galaxies, like autumn leaves swept into neat piles. Beyond our own galaxy, the spots of light we see through telescopes are entire galaxies. So the argument is the same, only now we consider galaxies rather than individual stars: surely the night sky should be as bright as an average galaxy – not quite as bright as the surface of a star, but still blinding?

Well, no. And, as we will see, the reason why not turns out to be one of the most profound truths about our universe that we have ever discovered. But in order to resolve the paradox satisfactorily, we must first see how it evolved through history.

An infinity of stars

Given how long astronomers have been aware of this paradox, it is somewhat surprising that it was as recently as the 1950s that it was attributed to, and named after, Heinrich Wilhelm Olbers, a nineteenth-century physician and amateur astronomer from Bremen in Germany. In fact, few astronomers even seemed interested in it until then.

In 1952 the great Anglo-Austrian cosmologist Hermann Bondi published an influential textbook in which the term 'Olbers' Paradox' was coined for the first time. But as we shall see, the attribution was misplaced, for Olbers was not the first to pose the problem, nor was his contribution to its resolution particularly original or enlightening. A century before him Edmond Halley had already stated it, and a century before *him* Johannes Kepler had posed it in 1610. And even he wasn't the first to record it: for that, we have to go back to 1576 and the very first English translation of *De revolutionibus*, the great work of Copernicus, written a few decades earlier.

Any account of the history of astronomy begins with the same few key individuals in the leading roles. First up is Ptolemy, the second-century Greek who, despite writing one of the most important scientific textbooks in history (known as the *Almagest*), believed erroneously that the Sun revolved around the Earth. He developed a model of the Universe with the Earth at its centre that held sway among astronomers the world over for a thousand years and more. Then there was Copernicus, the Polish genius of the sixteenth century who, in overthrowing Ptolemy's 'geocentric' idea and switching the roles of Sun and Earth, is regarded as the father of modern astronomy. And we must not forget Galileo, the first astronomer to point a telescope at the sky, in 1609, and in doing so to prove that the Copernican 'heliocentric' model was right: the Earth does indeed orbit the Sun, as do the other planets.

But Copernicus wasn't completely right. Though he was spot-on in removing the Earth from its privileged position at the centre of the Universe, he was wrong in simply replacing it with the Sun, believing that the solar system *was* the Universe. In *De revolutionibus*, regarded as one of the texts responsible for the scientific revolution in Europe, he shows an iconic diagram of the solar system. In it, the Earth is shown correctly to be the third planet out from the Sun after Mercury and Venus, with the Moon the only object in the heavens to truly orbit it; then come Mars, Jupiter and Saturn. All correct so far (the outer planets beyond Saturn were yet to be discovered) – but then Copernicus did something very interesting. He placed *all* the stars in a fixed outer orbit around the Sun. So he has the Sun truly at the centre of the whole universe, not just its system of planets.

We, of course, know now that our Sun does not have this special position. We know that in fact the Sun sits on an outer arm of an average spiral galaxy in a nondescript part of the Universe. We know, with the benefit of centuries of ever more comprehensive and accurate astronomical data, leading to our current understanding in modern cosmology, that the Universe has no centre at all, and may indeed extend out in all directions for ever. But of course Copernicus, working before the invention of the telescope, could not have known any of this.

It would take a relatively unknown astronomer in the sleepy English market town of Wallingford near Oxford to make the next big leap forward. His name was Thomas Digges and he was born in 1546, a few years after Copernicus died. His father, Leonard

Digges, was also a scientist – credited with the invention of the theodolite, an instrument used (these days mainly by surveyors) to measure vertical and horizontal angles very precisely. In 1576 Thomas published a new edition of his father's hugely popular almanac, *A Prognostication Everlasting*, adding new material to it in the form of several appendices. By far the most important of these was the very first English translation of Copernicus' great work. It is fascinating to think that this was tacked on to a book on astronomy that had itself not yet taken the new Copernican theory into account. However, Thomas Digges did more than just promote and publicize this still controversial view of the Universe, important though that was. He took it further in a way that is, for me, just as important an advance in astronomy as that of Copernicus, but far less celebrated.

Digges modified Copernicus' famous picture of the solar system, with its outer layer of stars fixed in their spherical shell around the central sun, and released those stars from their confining orbits, scattering them out into the unbounded and limitless void beyond. He was thus the first astronomer seriously to entertain the notion of an infinite universe containing an infinite number of stars (although the Greek philosopher Democritus had also hinted at such an idea).

This was no mere guess on Digges' part. He had been persuaded of his new picture of the Universe by an event that took place in 1572. Like many astronomers around the world that year, he had been transfixed by a new bright star in the sky. Today we would identify this rare event as a supernova – an exploding star at the end of its life that, having used up all its nuclear fuel, dramatically collapses under its own weight. This process sends shock waves through the star, causing its outer layers to explode into space with one final cataclysmic release of energy. Indeed, so much energy is released in this final burst that it briefly outshines an entire galaxy. But in the

Figure 3.1. Three models of the Universe

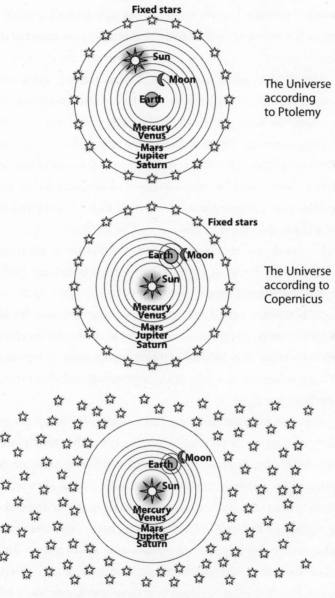

The Universe according to Ptolemy

The Universe according to Copernicus

The Universe according to Thomas Digges

sixteenth century, such astrophysical detail was not yet understood. Indeed, it was thought that, as the structure of the Universe beyond the Moon was fixed and unchanging, for an object in the sky to shine brightly for a while before eventually dimming again it would have to be very close to the Earth, and certainly within the orbit of the Moon.

Digges was one of several astronomers (others included the great Tycho Brahe) who calculated that the 1572 supernova had to be very far away. Since no daily shift in its position relative to other stars, known as 'parallax', was seen, astronomers were forced to deduce that it was further away than the Moon and planets. This was perplexing in the extreme; here was a heavenly body that had suddenly appeared from nowhere. It was referred to as a 'new star', and its appearance led Digges to conclude that the stars need not all be the same distance from us – maybe (as seems so obvious to us today) the brighter ones were nearer and the fainter ones much further away. Such an idea was quite revolutionary at the time.

As he contemplated the idea of an infinite space with an infinite number of stars, Digges was inevitably led to ask the crucial question: why is the night sky dark? For Digges, however, there was no paradox. He just assumed that the distant stars were simply too faint to contribute any light.

What Digges was missing was a vital mathematical calculation that would have shown the error in his reasoning about the darkness of the night sky. But that was to come later. In 1610 Johannes Kepler revisited the problem, arguing the reason it was dark at night was simply because the Universe was finite in extent: the darkness between the stars was the dark outer wall enclosing the Universe. Over a century after Kepler, another astronomer, the Englishman Edmond Halley, looked at the problem again and came out in support of Digges' original solution: that the Universe is infinite, but that the distant stars are too faint to be seen.

It was a Swiss astronomer by the name of Jean-Philippe de Chéseaux who showed, a few years later, that this does not help resolve the problem. He proved, using some neat geometry, that if we imagine all the stars grouped into concentric shells around us, like the layers of an onion, extending out to infinity – and assuming that on average the stars are all of the same brightness* throughout the Universe (which we know is not the case, but is nevertheless an acceptable assumption to make for the purposes of this proof) – then, while the stars in the innermost shells will shine most brightly, the shells further out, which because they are larger in area contain more stars, have an overall brightness that is exactly equal to any of the inner shells. In other words, lots of more distant and therefore fainter stars contribute as much light in total as fewer, nearer and hence brighter ones. We seem to be back where we started, with Kepler's argument that the Universe cannot be infinite or the night sky would not be dark.

Enter Heinrich Olbers, who again posed the problem of the darkness of the night sky in a paper he published in 1823. He offered a different solution. He knew, thanks to de Chéseaux, that the faintness of distant stars does not help resolve the puzzle. Instead, he argued that space is filled with interstellar dust and gas that would block the light from the more distant stars (or, as we now understand, the galaxies). What he failed to realize was that, given enough time, even this material would slowly heat up, owing to the light it absorbed, and would therefore eventually shine with the same brightness as the stars (or galaxies) it obscured.

In any case, Olbers' posing of the problem and his proposed solution were pretty much completely ignored by other astronomers right up to the end of the nineteenth century. But perhaps Olbers can be forgiven his mistake. You see, until this point, not only did astronomers

---

*Of course, at a certain distance we will have extended out beyond the Milky Way and would have to think about galaxies rather than stars,

not know how far out the Universe extended, they didn't even have good evidence that stars clustered in galaxies and that our Milky Way was just one of billions of galaxies scattered over vast distances. This would all change in the first decades of the twentieth century, when one man gave science a new view on the nature of space and time.

Our expanding universe

In 1915 Einstein published his greatest work. It wasn't his famous equation, $E=mc^2$; nor was it his work on the nature of light, which won him the Nobel Prize. It is known as the General Theory of Relativity, and in it he describes how the force of gravity affects space and time. We learn at school a description of gravity that Isaac Newton gave us: that it is an invisible force acting to pull all objects together. Of course, this is quite right, and we live our lives under the influence of our planet's gravity holding us to its surface. Newton's law of gravity also explains how the Moon orbits the Earth and how its gravitational pull affects the ocean tides; and it explains how the Earth orbits the Sun, thus confirming Copernicus' heliocentric model of the solar system. Indeed, it is the predictions of Newton's law of gravity that were used by NASA scientists when they sent the Apollo missions to the Moon. We are in no doubt that this universal law works. But it is not completely precise.

Einstein's General Theory of Relativity describes gravity in a radically different, and far more accurate, way. It says that gravity isn't really a force as such – like an invisible rubber band that pulls all matter together – but rather a measure of the shape of space itself around all masses. Now, unless you have a background in physics, these words will no doubt mean little to you. Don't worry; when Einstein first published his theory it was said that just two other scientists in the world understood it. Today, it has been tested and prodded so rigorously that we are left in little doubt of its correctness.

Since our Universe is basically space with stuff in it, and all stuff is governed primarily by gravity, it quickly became clear to Einstein and others that it should be possible to use the General Theory of Relativity to describe the properties of the whole Universe. But Einstein soon encountered a serious problem. If, at a given time, all the galaxies in the Universe are stationary relative to each other, and assuming the Universe is finite in size, then their mutual gravitational attraction should cause them to begin to converge in on each other, setting in motion the eventual collapse of the Universe. The generally held view at the time was that the Universe, at the level of galaxies and larger, was static and unchanging – the idea of an evolving, dynamic and changing universe on the very largest scales seemed to be both alien and unnecessary. So when Einstein's General Relativity equations suggested that the Universe should be shrinking, he decided, rather than contemplating a radical rethink, to patch things up. He argued that in order to balance the inward pull of gravity there needed to be an opposing force of antigravity, known as the cosmic repulsion force, which would balance the gravitational attraction and keep all the galaxies apart and the Universe stable. What Einstein had suggested was a mathematical trick in order to reconcile his General Theory with the prevailing model of a static universe.

But then came something of a surprise. In 1922 a Russian cosmologist by the name of Aleksandr Friedmann came to a quite different conclusion. What if Einstein were wrong and there were no force of antigravity holding the Universe balanced and stable? He realized that this would not necessarily mean that it had to be collapsing in on itself under the pull of gravity. It might also mean that the Universe is doing the opposite and is expanding. How could this be? Surely, without a cosmological force of repulsion the Universe had to be shrinking, not growing? Well, here's how.

Imagine that something had set the Universe expanding in the first place – some initial explosion. The gravitational pull of all the matter in the Universe would then be trying to slow the expansion down. So, if there were no cosmic force of repulsion to balance the attraction of gravity, and if the Universe had started off expanding (for some reason), then it would have to be either expanding or contracting at the present time. What it could not be was static: poised between expansion and collapse. That option is unstable.

Here is a simple example to demonstrate this. Consider what happens to a ball on a smooth slope: if placed halfway up the slope it will always roll down. But if we were watching a movie of the ball on a slope and paused the film with it frozen halfway up (or down) the slope and asked someone to predict what the ball would do next when we resumed play, then if they thought about it they should answer that it could be either rolling up the slope (which corresponds to an expanding universe) or rolling down the slope (a collapsing universe), but it could not be standing still. Of course, the only way it could be rolling up the slope would be if it had deliberately been given an initial kick. In that case, its motion up the slope would always be slowing, and it would eventually come to a stop and begin to roll down.

No one, not even Einstein, was prepared to believe Friedmann's theory – not until experimental proof was found. This came just a few years later. The astronomer Edwin Hubble was the first to prove that other galaxies existed beyond the Milky Way. Until then, it was believed that the tiny smudges of light that could be seen through telescopes were clouds of dust, called nebulae, within our own galaxy. With his powerful telescope, Hubble found that these other galaxies were too far away to be part of the Milky Way and therefore had to be galaxies in their own right. Even more remarkable was his observation that the furthest away of these galaxies were flying away from us at a speed that depended on their distance from the Earth. This seemed to be happening in whichever direction he pointed his telescope. He had proved that Friedmann's idea of an expanding Universe was correct.

Hubble argued, correctly, that since the Universe is now expanding, then in the past it must have been smaller. So if we went back far enough in time we would reach a point when all the galaxies overlapped each other and the Universe would have been pretty cramped. Go back even further in time and all the matter gets more and more squashed together until we reach the moment of the Universe's birth, that great explosion which we now call the Big Bang (a term first coined by the astrophysicist Fred Hoyle in the 1950s).

It is worth stating here that a common misconception of the expansion of the Universe is that all the other galaxies are flying through space away from our own. This is wrong. In fact, it is the empty space in between the galaxies that is stretching. It's also worth mentioning another interesting point – namely that our near-neighbour galaxy Andromeda is actually moving towards us! According to current estimates of the expansion rate of the Universe, it should be moving away from us at a speed of 50 kilometres per second. Instead it is moving towards us at 300 kilometres per second! The reason for this is that, just as stars are not evenly spread out through galaxies, so the galaxies themselves are not regularly distributed across the Universe. What Hubble had been observing was very distant galaxies moving away from us, rather than movement among those that make up our local cluster.

The speed at which our galaxy and Andromeda are moving together is equivalent to circumnavigating the globe in two minutes, or crossing the distance between the Earth and the Sun in under a week. In fact, Andromeda and the Milky Way are on a collision course, but at the current rate it will take several billion years for the two galaxies to merge.

One final point to make about the expansion of the Universe is that it appears that the rate at which it is expanding is increasing. It seems that, rather than gravity slowing down the expansion, something even stronger is pushing galaxies apart at an ever-increasing rate. It would appear that some mysterious force of antigravity, dubbed 'dark energy' for want of a better name, is at work. And so Einstein's idea of a force of cosmic repulsion was not so crazy after all – but now, instead of holding the Universe stable, it seems to be driving it apart.

Cosmologists now believe that while the Universe has indeed been expanding since its birth at the Big Bang almost 14 billion years ago, for the first 7 billion years the rate of expansion was slowing down, thanks to the gravitational pull of all the matter it contains – and then, during the second 7 billion years, that matter (the galaxies) became so spread out that gravity lost its grip. At this point dark energy started to take over, stretching space out faster and faster. What this means is that we now think that the Universe will never collapse back into itself again, in what is called the 'Big Crunch' (as was thought possible all the way up until 1998, when the acceleration in expansion was discovered), but will instead die a 'heat death' as everything moves away from everything else for ever – a somewhat depressing thought, I suppose. Not that we'll be around to worry about it, of course.

Proof of the Big Bang

Understanding that the Universe is expanding is actually enough to enable us to solve Olbers' Paradox – but let's go one important step further and prove that it is expanding because there must have been a Big Bang. Apart from the irrefutable evidence coming from the expansion of space, the Big Bang theory is today also supported by two other crucial pieces of evidence. The first concerns the relative proportions of different chemical elements in the Universe – what is known as the 'elemental abundances'. It is a fact that most of the atoms in existence are hydrogen and helium, the lightest two elements, with just a tiny amount of matter consisting of all the rest (oxygen, iron, nitrogen, carbon, and so on) put together. The only way to explain this satisfactorily involves a universe that was initially hot and dense, and then rapidly cooled as it expanded.

At the moment of the Big Bang, long before stars and galaxies had had a chance to form, all the matter in the Universe was squeezed together and there was no empty space. Almost immediately (much less than a second) after the Big Bang, subatomic particles began to form and, as the Universe expanded and cooled, these particles were able to stick together to make atoms. The conditions of temperature and pressure had to be just right for these atoms to form. If the temperature had been too high then the atoms would not have been able to remain intact and would have instead been smashed apart in the hectic maelstrom of high-speed particles and radiation. On the other hand, once the Universe had expanded a little more, the temperature and pressure had dropped too low to enable the atoms of hydrogen and helium to be squeezed together to form any other (heavier) elements. This is why mainly hydrogen and helium formed in the early Universe, a process that would have happened in the first few minutes after the Big Bang. Almost all the other elements had to wait until they could be cooked inside stars, where the conditions of ultra-high temperature and pressure were again available and where the process of thermonuclear fusion enables lighter atoms to be squeezed together to make heavier ones.

So the Big Bang theory is the only option for predicting the correct proportions of hydrogen and helium observed by astronomers today.

The other piece of evidence in support of the Big Bang was, just like the expansion of the Universe, predicted theoretically before it was confirmed experimentally. We know now that most of the photons travelling around in space are not starlight at all, because the Universe is awash with ancient light that has been around since before any stars or galaxies had even formed. Less than a million years after the Big Bang took place, the very first atoms began to form. At that moment space became transparent to light and radiation was free to travel vast distances. This light, the glow of that first universal dawn, has since then been stretched as the space it moves through has expanded. It was calculated that this light would have such a long wavelength by now that it would be beyond the visible spectrum. In fact, it would be in the microwave region. This is why it is called 'cosmic microwave radiation'.

This radiation, permeating the whole universe, can be picked up by radio telescopes as a faint signal from deep space. This was done for the first time in the 1960s and has been repeated many times since with ever-increasing sensitivity. Incredible as it may seem, we are able to hear the hiss of these faint waves as they are picked up by our radios and television sets.

So, the fact that our Universe had a beginning is no longer in doubt. Our three pieces of evidence are: the background radiation (the afterglow of the Big Bang – at just the right wavelength); the relative proportions of different elements; and the expansion of space that we see so clearly through our telescopes. All three point to this moment of creation.

Now, at last, we can finally lay Olbers' Paradox to rest.

The final solution

Let us recap. The reason the night sky is dark is not that the Universe is finite in size; for all we know, it may go on for ever. It is not that the distant stars are too faint; the further out we look, the more star-filled galaxies there should be, contributing their cumulative light to brightening up the gaps we can see between the stars of our own galaxy as we look out into space. Nor is it that the light from the furthest reaches of space is blocked from us by dust and gas that absorb it; given enough time, this intervening matter will also glow as it gradually absorbs the light energy it is blocking. No, the real reason for the darkness of space is more simple and profound than any of these suggested explanations. The night sky is dark because the Universe had a beginning.

Light travels at the mind-boggling speed of over 1 billion kilometres per hour. That's equivalent to travelling all the way round the Earth seven times in a single second. This speed is the cosmic limit for our universe. Nothing can go faster than light. It is not so much that light is special as that the speed itself is part of the fabric of our space and time. Light doesn't weigh anything, and this is what enables it to travel at the cosmic speed limit. Einstein showed this beautifully in the first of his theories of relativity, known as Special Relativity (which we will meet again in future chapters), in 1905 – and yes, if you must know, this is the theory that leads to the $E=mc^2$ relation.

And yet, on the cosmic scale the speed of light is not so impressive. For the distances that separate us from the stars in our galaxy, never mind the distances between galaxies, are so vast that the light from even our closest neighbours takes years to reach us.

It is the very *finiteness* of the speed of light that helps us resolve Olbers' Paradox. Since the Universe is nearly 14 billion years old, we can only ever see those galaxies that are close enough to us for their light to have had time to reach us. The expansion of space complicates matters, of course. A galaxy that we say is 10 billion light-years away from us is one whose light has been travelling towards us for 10 billion years. But during that time the space between us and that galaxy has been stretching, so the true current distance to that galaxy is in fact several times as great as that. However, a galaxy twice as far away as this one is out of our range; its light is still in transit towards us and we cannot see it. So it cannot contribute any brightness to the night sky. We can see out into space only as far as the age of the Universe allows us to.

What we can see in the sky, therefore, is just a tiny fraction of the whole cosmos. We call this the 'visible universe' and we cannot, even with the most powerful telescopes, see beyond this horizon in space. And this is because it is also a horizon in *time*. The further out we look, the further back in time we are looking; what we are seeing is the light that left its origin billions of years ago, and so we see it for what it was, not what it is. The edge of the visible universe is to us therefore also the earliest moment in time. And here is that final subtlety regarding the expansion of space. Even if an infinite *static* (not expanding) universe had suddenly popped into existence 14 billion years ago, we would still not be able to see beyond 14 billion light-years. So it is not the expansion itself that stops us seeing to infinity. For if we could wait long enough in a static universe then the light from ever-further galaxies would eventually reach us. It is just that beyond the edge of our visible universe the light will never outrun the expansion, like walking too slowly down an escalator that is going up.

In the previous chapter I made the point that in order to resolve Zeno's paradoxes we needed to appeal to rigorous science rather than abstract logic alone. But when it comes to Olbers' Paradox the first correct resolution was one based on intuitive logic rather than good science, and it came from the most unexpected of quarters: the nineteenth-century American author and poet Edgar Allan Poe.

The year before he died at the age of forty, Poe published what is widely regarded as his most important and influential work, an essay called *Eureka: A Prose Poem* (1848). Adapted from a lecture he gave and subtitled 'An Essay on the Material and Spiritual Universe', it is a remarkable piece of literature. It has no credibility as an actual work of science, being more to do with Poe's intuition about the laws of nature. In a sense, it is a treatise on cosmology in which Poe speculates on the origin of the Universe, its evolution and its end – and in which he relies on a mixture of logic and wild speculation rather than on valid scientifically backed ideas. For instance, he develops his own notions about how Newton's laws apply to the way planets are formed and how they spin, which are quite wrong. Nevertheless, buried in the essay is the following famous passage:

> Were the succession of stars endless, then the background of the sky would present us a uniform luminosity, like that displayed by the Galaxy – since there could be absolutely no point, in all that background, at which would not exist a star. The only mode, therefore, in which, under such a state of affairs, we could comprehend the voids, which our telescopes find in innumerable directions, would be by supposing the distance of the invisible background so immense that no ray from it has yet been able to reach us at all.

So there you have it. Olbers' Paradox was first resolved correctly not by a scientist but by a poet. Some historians argue that Poe's description is just guesswork and that it wasn't until a proper calculation was carried out by one of the nineteenth century's greatest scientists, Lord Kelvin, and actually published in 1901, that we can truly say the paradox was resolved. But Kelvin essentially provides the mathematical proof of Poe's idea. Whether we like it or not, Poe was right on the button.

So, to answer our initial question, *why does it get dark at night?* Because the Universe started with a Big Bang.

Final resolution and Big Bang proof

Scientists are often asked what proof they have that the Big Bang actually happened. They usually quote the three standard pieces of evidence

I discussed earlier. But isn't it so much easier, and in my view more persuasive, to turn Olbers' Paradox on its head? Rather than saying that the reason it gets dark at night is that the Universe must have had a beginning and that there has therefore not been enough time for light beyond a certain distance to reach us, why not try out the argument the other way round? For if anyone wants proof of the Big Bang they need only venture outside at night and ponder the darkness of space.

The real puzzle is that it took astronomers so long to figure this out.

FOUR

Maxwell's Demon

Is a perpetual motion machine possible?

IF YOU WERE to happen across a group of physicists and ask each of them to name what was in their view the most important idea in science, you might expect a wide range of answers: that everything is made of atoms, Darwin's theory of evolution, the structure of DNA, that the Universe began with a Big Bang. In fact, there's a good chance they would all opt for something called the Second Law of Thermodynamics. In this chapter we will explore this important idea and the paradox that, for over a hundred years, has stretched it to near breaking point.

The Paradox of Maxwell's Demon is a simple idea, and yet it has consumed many of the greatest names in science, and has even spawned whole new disciplines of research. This is all because it challenges that most sacrosanct of laws of nature: the Second Law of Thermodynamics, a simple yet profound statement about the transfer of heat and energy and how they can be used.

The Second Law of Thermodynamics says that if, for the sake of argument, you put a frozen chicken on top of a hot-water bottle – this is the example my family came up with when I tried to explain it to them – then you would expect the chicken to thaw a little and the hot-water bottle to cool down. You would *never* see heat going the other way, making the hot-water bottle hotter and the chicken colder. Heat always flows from warm bodies to cool bodies, never the other way round – and it does not stop flowing until an equilibrium is reached, so that there is no longer any temperature difference. Nothing controversial there, you might think.

Now let's look at the problem of Maxwell's Demon. Here is an outline of the initial idea to get us going. Imagine an insulated box containing only air. It is partitioned into two halves by a thick insulating wall. In the middle of this partition is a trapdoor that opens and closes very quickly when a molecule of air from either side gets close to it and allows that molecule through to the other side of the box. The pressure in the two sides remains the same, for if the number of molecules builds up on one side then it is much more likely that molecules from that side will come up to the trapdoor and be let through to equalize the pressure once more.

This process can go on indefinitely, and we would never get a temperature difference between the two sides. To explain this I need to define the concept of the 'temperature' of a gas down at the molecular level. Essentially, the faster the molecules bounce around, the hotter the gas will be. All gases, including the mixture we know as air, have trillions of molecules all moving around at random and at different speeds; some are faster, others slower. But their combined average speed will correspond to a certain temperature. Inside the box, some of the molecules passing through the partition between the two chambers will be fast-moving ones, others slower. On average, there should be as many faster molecules as slower ones crossing each way, and so no temperature difference builds up. If you're thinking that faster-moving molecules might be able to get through more often than slower ones then you could be right, but this does not affect the argument since as many faster ones should cross from the left to the right as will move back in the opposite direction.

Figure 4.1. Maxwell's box of air

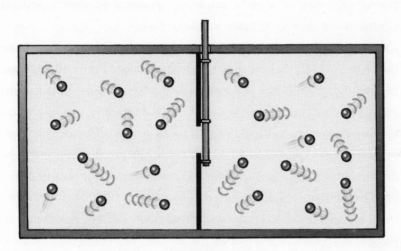

(a) Before

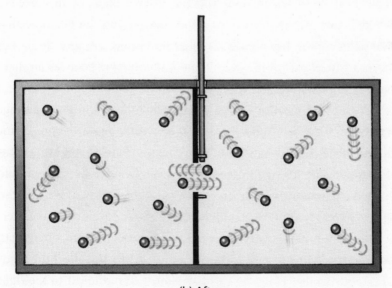

(b) After

If you're with me so far, I think I am now ready to release the demon.

Maxwell's demon is a hypothetical tiny creature that has such good eyesight it can see individual molecules of air and how fast they are moving. Rather than allowing the trapdoor to open and close randomly, we now let the demon control when it opens. Although it allows through just as many molecules as before, there is an additional factor to consider here: the demon's knowledge. For it allows only faster-moving molecules through from the left chamber to the right, and only slow-moving molecules from right to left. Given a gate-keeping demon armed with this knowledge, and with seemingly no extra effort or expenditure of energy (remember that earlier the trap-door was opening and closing randomly anyway), we find the outcome is now completely different.

It is tempting at this point to draw parallels with the role the game show host's knowledge plays in the Monty Hall Paradox, which we explored in Chapter 1. Don't fall into that trap. The fact that the game show host knew which door hid the prize influenced the way we worked out probabilities, nothing more. The knowledge that Maxwell's demon has plays a far more important role and, as we will see, is a critical part of the overall physical process that we are going to need to unpick to resolve this paradox.

With the demon in charge of the trapdoor, gradually the right-hand side of the box builds up with faster-moving molecules and gets hotter, while the left-hand side accumulates slower molecules and is therefore cooler. Using the demon's knowledge alone we seem to have created a temperature difference between the two halves, in violation of the Second Law of Thermodynamics.

So, with nothing more than information the action of Maxwell's demon seemed to reverse a process governed by the Second Law of Thermodynamics. How can this be? Many great minds over a period of more than a century have wrestled with this paradox. You are about

Figure 4.2. Maxwell's demon

(a) Before

(b) After

to find out how we resolve it – after all, like all the other apparent paradoxes in this book, it *can* be resolved, and the Second Law rescued.

The reason the topic has remained so fascinating is because of its connection to perpetual motion machines: devices that seem to be able to go on working indefinitely without consuming any energy. For if Maxwell's demon can violate the Second Law it should be possible to build a mechanical device that can do the same thing. Later on in the chapter I will look at a few types of such machines. By that time, I hope, you should not need much convincing of their impossibility.

Unwinding, mixing and rolling down hills

There are, in total, four laws of thermodynamics, which are all to do with how heat and energy can be transformed into each other, but none is as important as the Second Law. It has always amused me that one of the most important laws in the whole of physics cannot even make it to the number one spot on the list of thermodynamics laws.

The First Law of Thermodynamics is straightforward and states that energy can be converted from one form to another, but cannot be created or destroyed. It is usually formulated a little more technically than this: *that the change in the internal energy of a system is equal to the amount of heat supplied to the system, minus the amount of work performed by the system on its surroundings.* What this means is essentially that 'everything requires energy to do anything' – a car needs fuel, a computer needs electricity, we consume energy by just being alive, so we need food: these are all examples of how different forms of energy must be fed into a system for it to perform what is known as 'useful work'. The importance of the word 'useful' in this context is that it acknowledges that some forms of energy don't get put to productive use: for example, heat due to friction, or the noise of an engine, just dissipates into the surrounding environment. The First Law therefore simply lays the foundations for the more important Second Law. This states that everything wears out, cools down, unwinds, gets old and decays. It explains why sugar dissolves in hot water but does not undissolve. It also explains why an ice cube in a glass of water will inevitably melt, because heat is always transferred from the warmer water to the colder ice cube and never vice versa.

But why should this be so? If we were able to look at the world in terms of the collisions and interactions of individual atoms and molecules then we would not be able to tell in which direction time was flowing (by which I mean that if we were watching the process as a movie we would not know if it was running forwards or in reverse). This is because, down at the atomic scale, all physical processes are reversible. If a neutrino interacts with a neutron they will create in their place a proton and an electron that fly apart – but equally, a proton and electron could collide to make a neutron and a neutrino that move apart. The laws of physics allow both processes, forwards or backwards in time.

This is in sharp contrast with events that happen around us in everyday life, where we have no trouble deciding in which direction time is flowing. For instance, you never see smoke above a chimney converging on it and getting neatly sucked down inside it. Similarly, you cannot 'unstir' the sugar from a cup of coffee once it has been dissolved, and you never see a pile of ash in the fireplace 'unburn' to become a log of wood again. What is it that distinguishes these events from those on the scale of the atoms from which everything is made? How is it that most of the phenomena we see around us could never happen backwards? At what stage in going from atoms to chimney smoke, cups of coffee and logs of wood does a process become irreversible?

On closer examination, we see that it is not that the processes I have described above can *never* run in reverse, but rather that they are extremely unlikely to do so. It is perfectly possible within the laws of physics for dissolved sugar to 'undissolve' through stirring and reconstitute itself into a sugar cube. But if we ever saw this happening we would suspect some kind of conjuring trick – and rightly so, for the chances of its happening are so minute they can be ignored.

To help you understand the Second Law a little better I must introduce you to something called 'entropy'. It will feature quite largely in this chapter, so it is important to try to gain a clear idea of what it is. I should warn you now, though, that however carefully I try to explain it you will probably be left with a sense of it slipping from your grasp.

Entropy is a rather tricky concept to define because what it represents depends on the situation we are describing. Here are a couple of examples to illustrate it. One definition is that it is a measure of untidiness and disorder; how mixed up things are. An unshuffled pack of cards in which each suit is separate and arranged in ascending order (two, three, four . . . up to jack, queen, king, ace) is said to have low entropy. If we shuffle the cards a little the order is ruined and the entropy of the pack has increased. Now we can ask: what happens to the order of the cards if we then shuffle them further? The answer is obvious. It is overwhelmingly more likely that the cards become even more mixed up than it is that they return to their original ordered arrangement. So, with continued shuffling, entropy tends to increase. When the cards are completely mixed up entropy is said to be at its highest, and further shuffling cannot mix up the cards any more. The unshuffled pack is a unique arrangement of the cards, whereas there are very many ways for the cards to be mixed up, so it is overwhelmingly likely that shuffling will go in one direction: from ordered to mixed up – from low entropy to high entropy. This is the same irreversibility as in the case of a partially dissolved sugar cube, which on further stirring always continues to dissolve.

Figure 4.3. Entropy as disorder

The five cards on the left are arranged in a lower-entropy state
than the ones on the right.

We can see, therefore, that the Second Law of Thermodynamics is statistical in nature rather than relying on any specific property of the physical world. It is just overwhelmingly more likely that states of low entropy will evolve into ones of high entropy than the reverse.

To give you an idea of the probabilities involved, if you were to take a completely shuffled pack of cards then the chances of getting them ordered into the four suits, each consecutively numbered, through further shuffling is about as likely as it would be for you to win the National Lottery jackpot not once or twice but on nine consecutive draws!

On the other hand, entropy can also be thought of as a measure of something's ability to expend energy in order to carry out a task. In this case, the higher the ability to expend energy, the lower the state of entropy has to be. So, for example, a fully charged battery has low entropy, which increases as the battery is used. A clockwork toy has low entropy when wound up, which increases as it unwinds. When it has completely unwound, we can reset its entropy back to its original low value by expending energy ourselves in winding it up again.

The Second Law of Thermodynamics is basically a statement about entropy: namely, that it always goes up and never comes down unless additional energy is pumped in from outside. So, in the example of a clockwork toy, the Second Law is not violated when we wind it up because the system (the clockwork toy) is no longer isolated from its environment (us). The toy's entropy is being decreased but we are 'doing work' to wind it up, and our entropy is increasing by an amount that more than compensates for the drop in the toy's entropy. So overall, the total entropy of toy + us is increasing.

The Second Law thus also defines the direction in which time flows. You might think this is a rather trivial statement: surely time just flows from past to future. But 'flowing from past to future' is just the way we describe the process. To try to reach a more scientific definition let's imagine a lifeless universe so as to avoid our subjective distinction between the past (what we remember having already happened) and the future (what has yet to unfold). It then turns out that it is more meaningful and useful to say that time flows in the direction of increasing entropy, as we have now removed ourselves and the subjectivity of our brains from the process by defining the direction of time in terms of a physical process. This definition applies not just to individual systems, but to the entire Universe. So you can see that if someone were to come up with a situation where entropy in an *isolated* system was dropping, then you could say that time itself must have switched directions – and that is too weird even to contemplate (in this chapter anyway!).

Here is what the English astronomer Arthur Eddington had to say about the importance of the Second Law:

The law that entropy always increases – the Second Law of
Thermodynamics – holds, I think, the supreme position among

the laws of Nature . . . If your theory is found to be against the Second Law of Thermodynamics I can give you no hope; there is nothing for it but to collapse in deepest humiliation.

We sometimes see examples where it appears as though entropy is decreasing. For instance, a wristwatch is a highly ordered and complex system that is produced from a collection of bits of metal. Surely this is violating the Second Law? Well, no: in fact, this is just a more complicated version of the example of the clockwork toy. The watch-maker has put a certain amount of effort into making the watch, increasing his own entropy slightly. In addition, smelting the ores and machining the metals that are needed have produced a certain amount of waste heat that more than compensates for the small decrease in entropy achieved by the creation of the watch.

This is why Maxwell's demon presents us with such a puzzle. It seems to be able to achieve something akin to the watchmaker in lowering the entropy of the box by organizing the molecules of air, but without physically moving them itself. As a general rule, if it ever seems that entropy is decreasing, we always find that in fact the system under consideration is not isolated from its surroundings and that, by zooming out to look at the wider picture, we can see that entropy is increasing overall. We can view many processes that happen on Earth, from the evolution of life to the building of highly ordered and complex structures, as reducing the entropy on the surface of our planet. Everything from cars and cats to computers and cabbages has lower entropy than the raw materials it is made up from. Despite this, the Second Law is never broken. What we must not forget is that even the whole planet is not isolated from its surroundings. After all, almost all life on Earth, and hence all low-entropy structures, exists thanks to sunlight. When we consider the combined Earth + Sun system we see that the overall entropy is increasing because the radiation that the Sun pours out into space (only some of which is absorbed by the

Earth) means that its entropy is increasing by much more than the corresponding decrease on Earth, where the Sun's radiation is used to power life and, subsequently, all other complex low-entropy structures. A cabbage, for example, has taken in the Sun's energy via photosynthesis and has used this to help its growth, multiplying the number of highly organized cells it is made up from, thereby lowering its state of entropy.

So you can see how, over the years, scientists have been drawn again and again to the challenge of devising situations in which it would seem that the Second Law was being violated. Most notable of all was the nineteenth-century Scottish mathematical physicist James Clerk Maxwell, who famously figured out that light consists of fluctuating electric and magnetic fields. In a public lecture he delivered in 1867, he described his famous thought experiment involving the imaginary demon who is on a mission to defy the Second Law of Thermodynamics and who controls the trapdoor between the two chambers in the box. In controlling the door it acts like a valve that allows the energetic, 'hot' molecules of air to move only one way and the slow, 'cold' molecules to go only in the opposite direction. In doing so, it is sorting out the molecules, making one chamber hotter and the other cooler. This is in complete violation of the Second Law, since it does this without seemingly exerting any extra energy to open and close the door if, as I discussed earlier, it was opening and closing randomly anyway. And yet the entropy of the box as a whole appears to be dropping as the molecules within it are sorted into the two sides.

The one-way valve

So how do we go about resolving this paradox? Can Maxwell's demon lower the entropy of the box? And if so, how do we rescue the Second Law? Let me first approach this the way a physicist would: by stripping away any features of the problem that are not crucial to the argument

– in this case, by replacing our demon with a simple mechanical device that can do the same job. We can now ask whether a mechanical process exists that can replicate the demon's actions. And yes, the demon is, in a sense, acting like a one-way valve. We can therefore investigate whether such a valve can be used to create an imbalance between the two sides of the box by lowering the entropy, thereby providing us with a way to 'harvest' energy. Even before careful examination, this possibility should smell a little fishy. After all, if this were possible then surely the world's energy problems would be over. On that basis alone, the chances of its being possible would seem remote.

How can we be so sure, though, that one-way valves are unable to harvest energy from a state of equilibrium? Maybe the Second Law is not so sacrosanct. After all, everyone believed that Newton's law of gravity was perfect until Einstein came along and replaced it with a more accurate and radically different picture with his General Theory of Relativity. Could there exist a subtle loophole in the Second Law of Thermodynamics that just needs someone with sufficient brains, courage and imagination to come along and exploit it, replacing it with a better theory?

Unfortunately, the answer is no. Newton's law is based on his discovery of a mathematical formula that describes what is observed in nature, namely the way objects are attracted to each other depending on their masses and how far apart they are. Einstein showed that this formula wasn't wrong, just approximate, and that there was a deeper and more profound way of describing gravity in terms of the curvature of space and time – and, unfortunately, some much more complicated mathematics.

The Second Law of Thermodynamics is different. Although it may have originated from observations, it can be understood using pure statistics and logic, and it is now supported by a foundation stronger and more accurate than any observation. Indeed, Einstein himself wrote that it is 'the only physical theory of universal content which I am convinced will never be overthrown'.

So let us set up the simplified version of Maxwell's demon and see what happens. If you accept that any 'imbalance' between the two sides of the box that can slowly and spontaneously build up will correspond to a drop in entropy, then you will agree that we can replace the requirement of a temperature difference with one involving a pressure difference. After all, such a situation can also be used to perform useful work (as we shall see in a moment) and also corresponds to a lower-entropy state than when both sides are at equal pressure. Now, however, we would be dealing with a situation in which it is not that one chamber has fast-moving molecules while the other has slow ones, but that one side simply has more molecules than the other – hence the higher pressure. That is what we mean by pressure down at the molecular level: the number of molecules bashing against the walls of the chamber.

To see how an imbalance of pressure can be used to perform useful work, consider manually opening up the partition between the two chambers. If one contains air at a higher pressure, then air will rush through to the other side in order to equalize the pressure (with a corresponding increase in entropy). This flow of air could be used to perform useful work. For instance, it could drive a wind turbine and generate a little electricity. Clearly, then, creating such a pressure imbalance would be akin to storing energy, like winding up a clock-work toy or charging up a battery. For this to happen spontaneously would constitute a violation of the Second Law.

The simplest type of one-way valve we might use to do this is a swing door in the partition that opens in only one direction, allowing through molecules of air from the left side as they slam into it, forcing it to open, but spring-loaded so that it snaps shut again very quickly once they are through. Molecules hitting it from the right would just push it more tightly shut. Unfortunately, such a device would not even begin to work, for as soon as the slightest pressure difference built up between the two chambers, the pressure of the molecules hitting the door from the left would not be able to overcome the higher pressure on the right keeping it closed.

You might be thinking at this point that this device will only fail once the pressure difference is such that the higher-pressure right-hand chamber – the one with air molecules pushing against the door to keep it shut – stops those fastest-moving molecules on the left from forcing their way through. Surely the process can at least get started, with a little pressure difference building up as the first few fast-moving molecules are allowed through; and even this would violate the Second Law. Releasing even this smallest amount of pressure from the right-hand side to drive a turbine could generate a tiny amount of electricity. Since this process could be repeated again and again with more and more electricity being produced as a result, you can see how we would get into ever deeper trouble. We need to understand why *no* pressure difference at all can build up, or the Second Law is in trouble.

So far, we have been assuming that individual air molecules can kick open a trapdoor made up of trillions of molecules (of whatever material it consists of). In reality, if we are zooming down to the molecular scale then we must do so for the door too. And down at this level, the molecules of the trapdoor are also vibrating and jiggling about randomly. Even a single fast-moving molecule from the left-hand chamber that hits them to open the door will have given up some of its energy to the door molecules, making them vibrate a little more and, in the process, causing the door to open and shut randomly just enough to allow a molecule through the wrong way. Of course, there won't be precisely a one-to-one exchange, but with many molecules bombarding it from both sides, the trapdoor will be constantly vibrating down at its own molecular level and will never work as a one-way valve.

The same argument applies if we consider the possible build-up of a difference in temperature rather than pressure. Heat is essentially nothing more than the vibration of molecules and can be transferred through collisions between them, so this will apply to the molecules of the trapdoor just as much as to those of the air. Therefore, every time a fast-moving molecule from the left side hits the door and opens it, it transfers some energy to the door molecules, causing them to jiggle around more. This energy (or heat) will then simply get transferred back to the remaining air molecules in the left-hand chamber. So part of the energy of the fast-moving molecule finds its way back to the chamber it just came from. Any excess energy it takes with it to the right-hand chamber will ultimately be given up during its constant bombardment of the trapdoor from the right – energy which ultimately also gets transferred back to the left. This way just as many fast-moving molecules remain on the left as the right.

The lesson is this: the trapdoor acting as a one-way valve that reacts to molecules from one side only cannot itself be separated out from the process of energy transfer. If it is sensitive enough to react to individual molecules, then it is sensitive enough to be affected by them, and so can never act as an insulator between the chambers.

But the demon is cleverer than that . . .

I would like to introduce you to a Hungarian scientist and inventor by the name of Leo Szilárd. During an intense period of effort between 1928 and 1932, while in his early thirties, Szilárd invented some of the most important machines in history, all still used in scientific research today: the linear particle accelerator in 1928, the electron microscope in 1931 and the cyclotron in 1932. Incredibly, in all three cases, he never bothered to publish his work, patent his ideas or even build prototypes of the machines. All three inventions were subsequently developed by others who built on Szilárd's work. Indeed, two of them earned Nobel Prizes for other physicists (the American Ernest Lawrence, for development of the cyclotron, and the German Ernst Ruska, for building the first electron microscope).

It was during this period of intense creativity, in 1929, that Szilárd published a key paper that was to create something of a stir. It was entitled 'On the Reduction of Entropy in a Thermodynamic System by the Interference of an Intelligent Being', and in it he proposed a version of Maxwell's demon that has since become known as Szilárd's engine. But in his version it was no mere physical process that was at the heart of the paradox. Instead, he argued that it was indeed the demon's intelligence and knowledge about the state of molecules that made all the difference, just as Maxwell had feared. The paradox could not be resolved with a mechanical device, however ingenious.

Let me remind you of the paradox. Asking for the random bumping around of the air molecules to create, spontaneously and unaided, an imbalance in temperature or pressure between the two chambers simply won't work, no matter how cleverly the one-way valve or trapdoor operates – there always needs to be some external help. What is remarkable is that this help can, it seems, be in the form of simple information.

It appears we're back to where we started: trying to accommodate an abstract concept like information, possibly even the necessity of a sentient being, into the unthinking, mindless, statistical world of physical laws. Are we finally forced to concede that the Second Law of Thermodynamics holds only in a lifeless universe? That there is something magical about life that cannot be encompassed within physics? On the contrary, Szilárd's resolution was a brilliant affirmation of the universality of the Second Law and the notion of increasing entropy.

Imagine there are one hundred molecules in the box, with fifty in each chamber distributed randomly, such that there are as many fast as slow ones on each side, giving both compartments the same average temperature. (Of course, in reality there would be trillions of molecules, but let's keep things simple.) The demon transfers across, by careful control of when the partition is opened, the twenty-five faster molecules from one side and the twenty-five slower molecules from the other. This requires it to open the partition fifty times. You might think that the energy it expends in opening and closing the partition, however small an amount is required, is the way the demon pays the price of lowering the entropy. Such external energy would be the equivalent of winding up a clockwork toy – that is, reducing the entropy of something by carrying out work that has required an earlier increase in entropy from somewhere else. But if the demon has no information about the state of the molecules (by which I mean it cannot tell which are fast moving and which are slow) and simply opens the partition fifty times at random, allowing half the molecules from the left chamber to pass through to the right and half those on the right to go in the opposite direction, then the two sides would stay, on average, at the same temperature since as many fast molecules as slow molecules will be passed over in both directions. So, with no information, or having that information but choosing not to use it, we will not see any reduction in entropy. And yet the demon has still used up the same amount of energy opening and closing the partition fifty times. Clearly, the effort required to operate the partition need not have anything to do with the sorting process.

Szilárd's insight was to show how information fits in. He argued that the demon must consume energy not in operating the partition but in the act of measuring the molecules' speeds. Thus, gaining information always comes at an energy cost, energy that is expended in order to organize information in the demon's brain. At root, then, information is no more than an ordered state in the brain, or in the memory banks of a computer – namely, a low-entropy state. The more information we hold, the more structured and organized our brains are and the lower their entropy.

This low-entropy state of holding information gives us the ability to do useful work. So information is a little like a battery that stores potential energy, which can be used to lower entropy elsewhere.

Maxwell's demon will never be completely efficient, of course. It will use energy in order to gain information about the whereabouts and state (temperature) of all the molecules. It may then expend further energy in using that information to segregate the molecules. So energy used initially to gain information raises the entropy of the outside environment. Further energy used by the demon will have raised entropy outside even more.

In summary, then, we can think of a computer (or a brain) as a machine that can receive useful low-entropy energy, such as electricity (or food), and convert it into information – as opposed to useless, high-entropy energy such as the heat or noise produced by a motor. This information can then be used (or transferred) into a physical system to lower its entropy (by organizing that system, for example), giving it in turn the ability to carry out useful work. Since no step in this process will be 100 per cent efficient, a certain amount of heat will always be lost along the way. This hike in entropy associated with the dissipated waste heat gets added to the increase in entropy of the surrounding environment brought about by providing the demon with the sustenance it required to put in the effort of getting the information in the first place. Together, they more than compensate for the reduction in entropy as a result of the information processing. The Second Law is saved.

What does 'random' really mean anyway?

Let us take a more careful look at the Second Law and the issue of order and disorder, for we have yet to get to the bottom of what entropy really means. In the example of the shuffling of a deck of cards, there seemed to be little doubt that the entropy of an ordered pack, in which all cards were arranged in suits by ascending value, was low, and that a randomly shuffled pack had higher entropy. But what if the pack consisted of just two cards? Since there are now just two possible ways to arrange the cards, it doesn't make sense to distinguish between a less and a more ordered arrangement. What about three cards, say the two, three and four of hearts? Well, you might well say that the sequence 'two, three, four' is more ordered and hence of lower entropy than, say, the sequence 'four, two, three'. After all, the first sequence has them in ascending order. But what if the three cards were all twos, of hearts, diamonds and spades? Is one arrangement now any more ordered than the others? All that is different now is that the cards are defined by suit rather than value. Surely the way we label the cards cannot have a bearing on how much entropy there is? The sequence 'two of hearts, two of diamonds, two of spades' has no more, or less, entropy than the sequence 'two of diamonds, two of hearts, two of spades'.

It would seem that our definition of entropy as the amount of disorder is somewhat lacking because our definition of disorder is too narrow. It is obvious what we mean in some cases but not in others. Let me push this argument further. Here is a really bad card trick that demonstrates what I mean. I take an ordered pack of cards and shuffle it to reveal to you that the cards are now well and truly mixed up. Now, watch this, I say. I carry out what looks for all intents and purposes like a further normal shuffle. But, I claim, I have now placed the cards in a very special arrangement. This is an impressive claim, since it looked as though I was carrying out a similar shuffling action to the one that initially mixed up the pack. I turn the pack over and spread it out on the table. To your surprise and ill-concealed disappointment the cards look just as randomly jumbled up as they did earlier. This is certainly not what you would call a 'special arrangement', you argue.

Ah, but it is. You see, I could now bet you any money you could not take another pack of cards and shuffle it to produce exactly the same ordering as mine. The chances of your being able to do this are of course just as remote as they would be if I had asked you to take a shuffled pack and get it back to being completely ordered through shuffling alone. And the chances of doing that are about one in a hundred million trillion trillion trillion trillion trillion trillion. Basically, don't bother trying. So, looking at it this way we see that my randomly ordered sequence of cards is just as 'special' as a new, unshuffled pack. What of entropy now, then? It seems we cannot claim that entropy has increased if we end up in just as unlikely an arrangement as we started with, however randomly mixed up it looks.

Actually, I am trying to pull a fast one here. There is of course something more special about the ordered pack than my 'special' arrangement of randomly distributed cards. It comes down to entropy being a measure of *randomness* rather than disorder. This might seem just a play on words, but does in fact give us a tighter definition of entropy. Technically, the term used to measure relative levels of 'specialness' is 'algorithmic randomness'.

The word 'algorithm' is used in computing to denote a sequence of instructions in a computer program, and algorithmic randomness is defined as the length of the shortest program that can instruct the computer to reproduce a given arrangement of the cards (or sequence of numbers). Thus, for the earlier example with just three cards, reproducing the 'two, three, four' arrangement clearly requires the instruction: 'arrange from smallest to largest', whereas that for the 'four, two, three' arrangement might go something like 'start with the largest number, then put in increasing size', in which case it is just as easy to spell it out: 'start with the four, then the two then the three'. Either way, these commands have slightly higher algorithmic randomness than the first one, and so the arrangement 'four, two, three' has slightly higher entropy than 'two, three, four'.

This becomes much clearer when we have the whole 52-card pack. It is relatively easy to instruct a computer to reproduce the ordered pack: 'Start with the hearts and arrange cards in ascending order, ace high, then do the same for diamonds, clubs and spades.' But how would you program a computer to reproduce my special arrangement of shuffled cards? There is now really no shortcut and the instructions may have to be laid out explicitly, step by step: 'start with the king of clubs, followed by the two of diamonds then the seven of hearts [and so on]'. If the deck is not maximally disordered, there may be short sequences of unshuffled cards in which the original order has been preserved and which provides a saving on program length – for instance if the two, three, four, five and six of spades are still together, then it is easier to instruct the computer to 'start with the two of spades and arrange in ascending order for next four in same suit' than to spell out each of the four cards.

Figure 4.4. Entropy as randomness

The five cards on the left are arranged in a lower-entropy state than the ones on the right *because their order requires less information to describe,* not because it is more 'special'.

Talking about the length of computer programs may not mean much to you, and we can in fact dispense with that way of defining algorithmic randomness. Since our brains, just like the brain of Maxwell's demon, are, at their most basic, no more than computers running instructions, we can replace the notion of a computer algorithm with our memorizing ability. If I were to present you with a randomly shuffled deck of cards and then asked you to arrange them into suits and ascending order, this instruction is so special and simple that you could do it easily. (Note that I am allowing you to turn the cards face up and sort them properly rather than rely on blind, random shuffling to achieve this by chance.) But if I were to ask you to arrange your pack in the same order as my 'special' arrangement, which I had arrived at by random shuffling, you would probably find it near-impossible to memorize the order of the cards before you attempted to replicate it with the pack in your hands. Basically, you need a lot more information to reproduce the arrangement of the pack now than you did before. And the more information you have about a system, the more you will be able to order it and lower its entropy.

Perpetual motion machines

Throughout history, many enterprising souls have attempted to invent a perpetual motion machine that could keep running and producing useful work indefinitely; or, to put it more simply, that would produce more energy than it consumed, even if just to keep itself going. This is impossible.

Let me first make clear that we should always be careful when we say that something in science is impossible. After all, the statistical nature of the Second Law of Thermodynamics has taught us that it is not completely impossible for an ice cube to form spontaneously in a glass of warm water. However, this is so exceedingly unlikely that you would have to wait far longer than the age of the Universe for it to happen – so we can rule it out. Usually, when we say something is impossible we mean 'impossible according to our current under-standing of the workings of nature and the currently accepted theories of physics'. Of course, we may be wrong, and it is that glimmer of hope that keeps some inventors trying to design ever more ingenious perpetual motion devices.

Such machines can be categorized into two main types. Perpetual motion machines of the first kind are ones that violate the First Law of Thermodynamics since they produce work without any input of energy. The First Law of Thermodynamics is a statement of conserva-tion of energy, which means that, in any closed system, you cannot create new energy. Any machine that claims to produce energy from nowhere will be of this type.

A perpetual motion machine of the second kind is one that, while not violating the First Law, goes against the Second Law of Thermo-dynamics by converting thermal energy into mechanical work in a way that causes entropy to decrease. The subtlety here is that this would happen without any corresponding opportunity for entropy to increase elsewhere to balance things out. As I mentioned earlier, one way of stating the Second Law is that heat flows from a hotter place to a cooler one. As it does so, entropy rises, but useful mechanical work can be extracted from this process to lower entropy elsewhere provided that the fall in entropy is not greater than the rise in entropy caused by the heat transfer. A machine that can extract energy from a hot object without any accompanying flow of heat to a cooler place, such as the Maxwell's demon set-up, is one such attempt at a perpetual motion device.

There are of course many devices that comply with both laws of thermodynamics by getting their energy from some obscure and subtle external source, such as air pressure, or humidity, or ocean currents. These are not perpetual motion machines since they violate

Figure 4.5. Two simple perpetual motion machines

(a)   (b)

(a) *The 'overbalanced' wheel.* This idea for a perpetual motion machine goes as far back as eighth-century India. Many elaborate designs have been suggested, but all are based on the same principle, and all fail for the same reason. In the version shown here, the balls on the right-hand side (between 'quarter past' and 'half past') roll to the outside and, by virtue of being further from the centre, will have greater torque (turning force) than if they were at the middle. This suggests that they will win over the balls on the left-hand side and force the wheel round clockwise continuously once it is set slowly spinning. In practice, there are always more balls on the left acting against the spin than balls with the greater torque on the right keeping it going, and it inevitably slows down to a stop.

(b) *The magnetic motor.* Here, the idea is that the central magnet is shielded from the outer ones arranged in a ring, apart from the two gaps where each of its north and south poles feel the magnetic field of the surrounding magnets. At the top, its south pole is attracted to the north pole in the inner ring while underneath it, its north pole is repelled. Both act to turn the magnet clockwise . . . for ever. The flaw here lies in a confusion over how magnetic fields work. In fact, there is no magnetic field at all inside the circle: it is cancelled out by symmetry, and the inner magnet feels no turning force at all.

no laws of physics. You just have to figure out what power source is keeping them going.

Some devices, such as those involving spinning wheels or swinging pendulums, look on first inspection as though they might keep going for ever without any power source. This is not the case. They are simply very efficient at not leaking away the initial energy that got them started, which they will of course need to have had. In fact, they will all eventually slow down, since no machine can be 100 per cent efficient and there will always be some form of damping, because of friction in air or in any moving parts, however well oiled they are.

In principle, therefore, a perpetual motion machine is possible provided no energy is lost to its surroundings. Of course, any attempt to extract energy from such a device will bring it to a stop.

Maxwell's demon and quantum mechanics

The debate over Maxwell's demon did not end with the work of Szilárd. Physicists today have chased the demon all the way down to the quantum realm and the strange rules that operate at atomic scales. In quantum mechanics, as soon as we begin to discuss the idea of measuring the positions and speeds of individual molecules, we come up against a fundamental issue to do with just how much information we can acquire. It is known as Heisenberg's Uncertainty Principle and states that we can never know exactly both *where* a particle (or air molecule) is *and at the same time* exactly how fast it is moving; there is always a kind of fuzziness. And it is this fuzziness, many argue, that is ultimately needed to preserve the Second Law of Thermodynamics.

It would seem that the quantum world has become the last bastion of hope for those still holding on to the dream of a perpetual motion machine. For a number of years now, suggestions have been made that this might be possible using what is referred to as vacuum energy, or zero point energy. Because of the fuzziness of the quantum world, nothing can be said to be truly at rest, and every molecule, atom or subatomic particle will always have a minimum amount of energy, even if cooled to absolute zero: this is called its 'zero point energy'. This applies even to the vacuum of empty space: according to quantum physics, the whole universe is full of this 'vacuum energy', which many think we should somehow be able to harvest and use. However, such approaches run into the same difficulties as we encountered with the chambers of air molecules. Vacuum energy is evenly spread out, so any methods of extracting it and putting it to use would require the expenditure of more energy than the amount we can get out. Evenly distributed vacuum energy cannot be freely harvested, just as a temperature difference between the two chambers of the box cannot be created without some help.

That help can come in the form of information – like that held in the mind of Maxwell's demon – but we still require energy to obtain that information in the first place; energy that has to be paid for by raising entropy elsewhere.

We can never defeat the Second Law of Thermodynamics. Always remember that.

Oh, I almost forgot: I mentioned at the start of the chapter that there were *four* laws of thermodynamics, but I didn't tell you what the other two were. Don't hold your breath: the Third Law of Thermodynamics states that the 'entropy of a perfect crystal drops to zero when the temperature of the crystal is at absolute zero'. As for the fourth law, the only interesting thing about it is that although it was added to the other three long after they were established, it is seen as more basic and fundamental and is therefore known not as the Fourth Law, but as the Zeroth Law, since it needs to come before the other three. It states that if two bodies are in thermodynamic equilibrium (the scientific way of saying 'at the same temperature') with a third body, then they will be in thermodynamic equilibrium with each other – hardly mind-blowing. This law was given the number 'zero' simply because the other, more important, laws were too established to be shifted up in number. That would have led to too much confusion – and that's the last thing we would want, right?

FIVE

The Pole in the Barn Paradox

How long is a piece of string? It depends on how fast it is moving...

Y OU WILL MOST likely not have heard of this paradox unless you are, or were once, a physics student, for it is one of a handful of famous text-book examples used to teach Einstein's theory of relativity which highlight certain weirder aspects of the theory's predictions regarding the nature of space and time. But this paradox is too delicious to be left to physicists to enjoy, and I am going to take some pleasure in sharing it with you. I warn you now, though: it is not one that can be stated first, without any prior knowledge of the physics involved, then explained away with a little bit of, in this case, relativity theory to clear things up. Here, you have to be introduced to some of the physics first, before I can even pose the paradox satisfactorily, let alone resolve it!

Nevertheless, I did promise at the outset that I would outline each paradox at the start of the chapter, so that you knew what you were letting yourself in for. Here it is, then – and please hold on to your scepticism, for it should serve you well as we dive headlong into Einstein's world.

A pole-vaulter runs, while holding a pole parallel to the ground, at very high speed – in order for this to work, we must assume that he can run at close to the speed of light! He approaches a barn that is of the same length as his pole. He knows this because he measured the pole against the side of the barn before he started off on his run-up. The barn's front and rear doors are wide open and he carries on running through it without slowing down. With no knowledge of relativity theory, we would assume that there will be a moment in time when the rear of the pole has just entered the barn while, at the same time, the front of the pole is just exiting.

This would be fine if the pole-vaulter were running at normal human speed. But he is not; he is running at close to the speed of light, and this is where all sorts of strange and wonderful physical effects, predicted by Einstein's theory of relativity, become manifest. One of these – the one crucial for this discussion – is that a fast-moving object will appear shorter in length than it does when standing still. Of course, you might be thinking, I can buy that; after all, it zooms past so fast that by the time I clock where the front of it is, the back has moved forward, giving the impression of something shorter. No, no, no. If only it were that simple.

If you fire a missile (measured earlier to be exactly 1 metre long) at close to the speed of light so that it travels along the length of a fixed measuring tape, and then take a snapshot of it in mid-flight, you will see that its length is now less than 1 metre – by how much depends on its speed: the closer to the speed of light it moves, the more squashed it will be. I will go into this more carefully later on, but for now let us return to the pole in the barn.

To restate: relativity theory tells us that if you were standing in the barn watching the pole-vaulter run through, then you would see the pole to be shorter than the length of the barn. The rear of the pole would enter the barn at some moment and only later would the front exit the other side; there would be a short period of time when the whole pole was inside the barn.

While very bizarre, this is still not yet a paradox, for there is another important lesson we learn from relativity theory. Indeed, it is the lesson that gives the theory its name: *all motion is relative.* This idea goes back long before Einstein and there is nothing really strange about it. Imagine you are on a train and another passenger walks past your seat in the direction of travel of the train. Since you and he are both moving with the train, he walks past you at the same walking pace he would do if the train were stationary. However, at that very moment the train is passing through a station and someone on the platform also sees the passenger walking through the train. To him the passenger is moving past at a speed that is the combination of his walking pace and the much faster speed of the train itself. So the question is this: how fast is the passenger moving? At the walking pace relative to you, or at that speed plus the train speed's relative to the observer on the platform?

We feel comfortable saying that the answer depends on the observer. Speed is not absolute, as it depends on the state of motion of the person measuring the speed. In a similar way, you could argue that while you are sitting on the train, the train itself can be considered stationary and the outside platform is moving in the opposite direction. This might seem like pushing the idea too far, for surely it is more correct to say that it is *really* the train that is moving. But consider this: what if the train were moving at 1,000 miles an hour (not realistic, I know) and travelling east to west? Imagine you are floating out in space; what do you see? Well, you would see the Earth spinning at about 1,000 miles per hour in the opposite direction to the train. This is what it takes for it to do one revolution on its axis every day. To you, the train is keeping pace with the rotation of the ground underneath it and is thus not moving – it's like watching someone running on a treadmill. Now, would you say it is the train or the planet that is moving? You see? All motion is relative.

Good. I assume you're convinced so far. So let us return to the pole in the barn. From the point of view of the pole-vaulter, even though he is pumping his legs at a far from realistic rate, he can still regard himself, and his pole, as stationary, and the barn as approaching him at nearly the speed of light. The theory of relativity here is very clear: to the pole-vaulter, the barn is moving and he would see it as shortened in length – much shorter, in fact, than his pole. So, to him, by the time the back end of the pole has gone through the entrance of the barn, the front will have long since exited. In fact, there was a period of time when the pole would have been sticking out of both ends of the barn.

Here, then, is the paradox: to you, watching the shortened pole enter the barn, the pole is shorter than the barn, and you could conceivably close both front and rear doors of the barn (with a suitable trigger) for a fraction of a second, enclosing the whole of the pole within it. But to the pole-vaulter, the pole is longer than the barn – far too long ever to be contained. Surely you cannot both be right? And yet, the correct answer is that *you are indeed both right*. This is the Pole in the Barn Paradox, and I shall spend the rest of this chapter not only resolving it, but explaining how and why relativity theory forces this awkward dilemma on us in the first place.

To solve this particular paradox we are going to have to delve bravely into Einstein's theory of relativity, and by following along the path he took over a century ago we will stride purposefully from one logical stepping stone to the next until we reach our destination.

I feel it is best to be completely honest here. I certainly do not plan to use any algebra or draw any technical graphs in order to teach you the basics of relativity, and I could in principle just jump to the resolution of the paradox in the hope that you are happy to trust me on this business of lengths getting shorter at ultra-high speeds. But then again, I could just be making this stuff up. So you

have a choice: you can skip to the end of the chapter where I
explain away the paradox if you (a) know something about the
Special Theory of Relativity already or (b) trust that if Albert
Einstein says so then that is good enough for you; or you can allow
me to lead you through the arguments carefully and gently. If you go
for the latter option it will be worth it in the long run, since the next
two chapters, on paradoxes involving the nature of time rather than
space, will also depend on what I explain here. And I promise I shall
do my very best to make it not only painless, but possibly even fun.
After all, special relativity is one of the most beautiful theories
in physics.

A lesson on the nature of light

By the late nineteenth century it was clear that light behaves like
a wave – just like sound, only moving much faster. There are two
important properties of waves that you need to be aware of in
order for the rest of this discussion to make sense. First, waves need
some medium to pass through; some 'stuff' that does the 'waving',
or vibrating. Consider how sound travels: when you speak to
someone standing next to you, the sound waves that get from
your mouth to their ear move through the air. It is the air
molecules that vibrate and carry the sound energy across. Likewise,
waves on the surface of the sea need the water, and the 'bump' that
travels along a length of rope when it is given a flick at one end needs
the rope.

   Clearly, without the medium to carry the wave along there would
be no wave. So it is understandable that nineteenth-century physicists
were convinced that light, having been defined as an electromagnetic
wave, also needed some medium to travel through. And since no
one had seen such a medium, they had to try to come up with an

experiment to detect it. It was labelled the luminiferous ('light-carrying') ether and a major effort was made to prove its existence. Of course, it had to have certain properties: for instance, it had to permeate the whole galaxy in order for distant starlight to reach us through the vacuum of space.

In 1887, at a college in Ohio, two American physicists, Albert Michelson and Edward Morley, conducted one of the most famous experiments in the history of science. They had devised a method for measuring very accurately the time it takes for a light beam to cover a fixed distance. But before I describe what they found, there is another property of waves that I need to mention, which is that *the speed at which waves travel does not depend on how fast their source is moving.*

Think of the noise of an approaching car. The sound waves will reach your ear before the car since they are travelling faster, but their speed is to do with how quickly the vibrating air molecules can transmit them. They do not reach you any faster by virtue of being 'pushed' along by the moving car. What happens instead is that as the car gets nearer to you the waves between it and you get squashed up to shorter wavelengths (or higher frequencies). This is called the Doppler effect, and is familiar to us as the change in pitch of approaching then receding ambulance sirens or the loud engines of racing cars passing on a track. So, while the *frequency* of a sound wave does depend on the speed of its source and whether it is approaching or moving away from us, the *speed of the wave itself* – the time it takes to reach us – does *not* change.

Crucially, however, the situation is completely different when considered from the point of view of the driver of the car. The sound of the engine will be carried through the air away from the car at the same speed in all directions. So, in the direction of travel of the car, the sound waves are moving away from it *more slowly* than they are moving away at right angles to it. This is because the speed at which the sound waves move ahead of the car is the difference between the speed of the waves in air and the speed of the car.

Michelson and Morley applied this principle to light waves. They devised an ingenious experiment – one which they were convinced would be the first to confirm and detect the existence of the luminiferous ether. They began by assuming that the Earth is moving through the ether as it orbits the Sun, which it does at about 100,000 kilometres per hour. In their laboratory experiment they measured, with incredible accuracy, the time it took two light beams to travel along two different paths of equal distance, one in the direction of the Earth's motion as it orbited the Sun and the other at right angles to it. Sitting in their laboratory on Earth and observing the speed of light, they were like the car driver who finds that sound waves leaving the car have different speeds depending on whether he measures them ahead of him or out to one side.

If the ether exists, argued Michelson and Morley, then the Earth has to be moving freely through it, and so light travelling in different directions would take different lengths of time to cover the same distance; for it would be moving, relative to the moving Earth, at different speeds in the two directions. Although the speed of light is 300,000 kilometres per second, which is ten thousand times as fast as the speed of the Earth in its orbit, their measuring apparatus, called an interferometer, was so accurate that it could pick up any difference in the journey time between the two light beams by the way their waves interfered with one another when they were combined at the end.

No such time difference was found.

Their experiment had yielded what is known in science as a 'null result' (and one that has since been repeatedly confirmed by many more precise experiments using laser beams). The world's physicists could not understand this outcome – in fact, they believed Michelson and Morley had made a mistake. How could the two light beams be moving at the same speed? What had happened to the 'all motion is relative' principle?

I know this all sounds a bit confusing, so let me put it as starkly as possible. Remember the example of the passenger walking through the train? The Michelson–Morley result was the equivalent of both you on board the moving train and the observer watching the train go by agreeing on how fast the passenger was moving! It sounds ridiculous, right? Surely, as I explained before, you see the passenger moving at walking pace while the platform observer sees him whizz past at train speed, plus a little more.

Just eight years before Michelson and Morley achieved their disturbing finding, Albert Einstein had been born in Ulm in Germany. That same year, 1879, Albert Michelson, working at a US Naval Observatory in Washington, had measured the speed of light to an accuracy of about one part in ten thousand. He wasn't the first to do this and would not be the last, but it would stand him in good stead when he and Morley conducted their famous experiment. As for the young Einstein, although of course he was

---

Figure 5.1. Einstein's early research

Can Einstein, flying at the speed of light, still see his reflection?

completely unaware of the astonishing result that Michelson and Morley announced to the world, he was nevertheless soon pondering the unusual properties of light himself by devising imaginary experiments. He asked himself whether, if he could fly at the speed of light while holding a mirror in front of him, he would still see his own reflection – for how could the light from his face ever reach the mirror if the mirror itself was always moving away at the speed of light? His years of contemplation culminated in 1905 when, still only in his mid-twenties, he published his Special Theory of Relativity. Suddenly, Michelson and Morley's result could be understood beautifully.

Until Einstein published his theory, physicists had either refused to believe Michelson and Morley's result or had tried to modify the laws of physics to accommodate it, but to no avail. They tried to argue that light was behaving as a stream of particles (since that would also explain the result), but the experiment was set up specifically to detect the wave nature of light by relying on the way the waves in the two beams of light overlapped as a means of measuring their precise arrival time. In any case, if light were made up of particles then that would also do away with the ether since particles would not require a medium to travel through.

All this changed in 1905. Einstein's whole theory rested on two ideas, which have become known as his two postulates of relativity. The first was an old one. It stated simply that all motion is indeed relative and that nothing can be said to be truly stationary. This means that there are no experiments we could perform that would tell us whether we were truly standing still or moving. The second postulate was the revolutionary one, although it sounds quite innocent at first. Einstein stated that light does indeed have the wavelike property that its speed is independent of the speed of its source (just like the sound waves from a moving car). Yet at the same time, and unlike sound waves, light does not require a medium to pass through; the luminiferous ether does not exist and light waves can move across truly empty space.

So far, so good; no paradox here – and nothing, you might think, in either of these innocuous postulates that you might have difficulty subscribing to. They certainly don't sound like statements that lead to a revolutionary view of space and time. But they do. Each postulate is, on its own, innocent. It is when the two are combined that we see how profound Einstein's ideas were.

Let us recap. Light reaching us from a source will travel at the same speed regardless of how fast the source is moving. This is just like other waves, such as sound, so no problems here. However, because light doesn't have a medium to travel through, with respect to which we can measure its speed, then no one has a privileged position in the Universe and we should all measure light to have the same speed (1 billion kilometres per hour) regardless of our own state of motion. This is where things get weird, so I will explain what it implies.

Consider two rockets travelling at high speed towards each other in space. If both the rockets' engines are turned off and they are just 'cruising' at constant speeds, then no one on board either rocket would be able to determine if they were both actually moving towards each other or whether one rocket was stationary and the other approaching it. In fact, there is no such thing as moving or stationary since motion must always be with reference to something else. So it is no good appealing to a nearby star or planet as a reference point, since who is to say that even that is stationary?

Now, an astronaut on board one of the rockets shines a light beam towards the other rocket and measures the speed of the light as it leaves his rocket. Since he can quite legitimately claim to be stationary, with the other rocket doing all the moving, he should see the light moving away from him at the usual billion kilometres per hour. At the same time, the astronaut aboard the other rocket can also legitimately claim to be stationary. He too measures the speed of the light reaching him at 1 billion kilometres per hour and states that this is not at all surprising since the beam's speed does not depend on how fast its source is approaching. This is exactly what we find. Paradoxically, both measure the light to have the same speed.

This is amazing, and goes quite against common sense. Both astronauts measure the same light beam to be travelling at the same speed, despite the fact that they are moving towards each other at nearly the speed of light themselves!

Before we go on, we can now answer Einstein's question involving the mirror. It does not matter how fast he flies, he will always see his reflection. This is because regardless of his speed he still sees light travelling at the same speed from his face to the mirror and back again, exactly as he would were he not moving at all. After all, who is to say he is flying so fast? All motion is relative, remember?

There is a price to pay for all this, and it is that we must overhaul our views of the nature of space and time. The only way light can travel at the same speed for all observers regardless of how fast they are moving with respect to each other is if *they measure distances and times differently*.

Shrinking distances

Before you complain that this is just some speculative 'theory', which might turn out to be wrong, I must stress that it has been studied and tested for a hundred years and we see its effects routinely. I can even vouch for it personally since I, like many physics students, performed a laboratory experiment while at university involving a type of subatomic particle called a muon (pronounced 'mew-on') that is produced by cosmic rays (high-energy particles from space that constantly bombard the upper atmosphere). The muons are created by collisions of the cosmic rays with air molecules and stream down to Earth. The laboratory experiment I conducted as a student involved capturing and counting these particles in a special detector. We know that muons exist for only a tiny fraction of a second before they disappear – something we measured carefully in the experiment. Generally, this lifetime is only about two microseconds, though some might live for a little longer, some for a little less.

The muons are so energetic that they travel towards the Earth at over 99 per cent of the speed of light. However, even at this speed it should still take them several muon lifetimes (which we can work out from their speed and the rough distance we know they have travelled) to cover the distance to the surface of the Earth. We should therefore only be detecting those few unusually long-lived ones that are able to complete the journey. Instead, we find nearly all the muons are comfortably able to reach and trigger our detector before disappearing. A possible argument might be that fast-moving muons live longer than stationary ones for some reason. However, Einstein would say that this cannot be the explanation since all motion is relative, so a moving muon is only 'moving' relative to us on the surface of the Earth.

Here then, at last, is the payoff. Consider how things would look from the muon's point of view. If it could speak, it would tell you that it is indeed travelling at over 99 per cent light speed – or rather, that the ground is coming up to meet it at over 99 per cent light speed. And it seems to have enough time to cover this distance. In fact, from its perspective, the time it takes to reach the ground is so short that it is well within its brief lifespan. This can mean only one thing: time must be running more slowly for the muon than it is for us on Earth. And this is indeed the case – but I will postpone going into further detail about this slowing down of time until the next chapter. For now, we have one final logical hurdle to cross. Consider this: first, you and the muon agree on the speed at which it is travelling (or, more precisely, the speed you are approaching each other); second, the muon says its journey takes less time than you think. So, to balance the books it must also say it is covering a shorter distance. That is, if, travelling at some fixed speed you both agree on, the muon is able to cover the distance in less time, then it must see that distance as shorter than you do.

This property of high-speed travel is known as length contraction. It states that, just as fast-moving objects look shorter than the same objects do when standing still, so distances to be travelled look shorter when viewed from the perspective of the fast-moving objects.

Galactic travel

This has an interesting consequence that is worth exploring briefly before we return to the Pole in the Barn Paradox. When introducing Olbers' Paradox in Chapter 3, I mentioned that our nearest neighbour stars are several light-years away from us. So even if we could travel at light speed it would still take us years to reach them. This is a rather depressing thought since it implies that we are trapped in our solar system and that the best we can realistically hope for is a visit to the other planets orbiting the Sun; getting any further would simply take too long. As for visiting more distant stars, let alone new galaxies 'far, far away', well, this seems out of the question – after all, even light would take thousands or millions of years to get there.

So, what if I told you that you could travel below the light-speed limit and still get to the other side of the Universe in the blink of an eye? Science fiction? No. The only thing stopping us is the fact that we don't have a rocket that can travel at near light speed, and we may never have. But let us assume we do. The way this works is exactly the same as in the example of the muon. Just as from the muon's perspective the distance to the ground is much shorter than we see it, so too a traveller on board a spaceship heading to a distant star at close to the speed of light would see the distance to be covered squashed up.

Imagine there was a solid rod, thousands of light-years in length, which joined the Earth to the destination star. Those on board the spaceship could argue that because all motion is relative it is not the ship that is travelling at near light speed, but the rod that is moving at the same speed in the opposite direction. From their perspective, they are stationary, watching the high-speed rod move past. They will therefore see its length shortened; and so it will not take as long to pass them – and it will not take them so long to reach their destination.

Relativity theory tells us that the closer you get to light speed, the more lengths are contracted. So, to give you an example, a distance of 100 light-years will appear to a traveller moving at 0.99 of the speed of light relative to its start and end points to be just 14 light-years. But to one travelling at 0.9999 of the speed of light, this distance would appear to be just 1 light-year (and hence the journey time would be just one year, since the spaceship is travelling at just about the same speed as light). And if the ship could nudge even closer to the speed of light, say 0.999999999 of the speed of light, then 100 light-years would be covered in less than two days.

Notice that we are not violating any laws of physics here. The closer you travel to the speed of light, the shorter the time it would take to reach your destination, and I hope you can now appreciate that this is not because you are going faster (after all, 0.999999999 of light speed is not much more than 0.999 of light speed in absolute terms) but because the closer you can get to the speed of light, the shorter the distance will appear to you; and the more the distance shrinks, the less time is required to cover it.

So is there a price to pay? For you on board the spaceship, covering the 'shrunk' distance means that little time will pass and the journey will be over very quickly. If it takes two days to cover 100 light-years then you will indeed be just two days older when you get there. But remember, in comparison with the passage of time on Earth your time is running much more slowly. As far as everyone else back on Earth is concerned, you are travelling at near light speed and having to cover a distance of 100 light-years, and so the journey will take 100 years (or marginally over, since you are moving at just below the speed of light). So your two 'spaceship days' correspond to 100 'Earth years'. Even worse, if you send a light signal back to Earth when you arrive, then it will take a further 100 years to get here. So the first indication that you've got there safely will not reach Earth until 200 years after you left.

The lesson here then is that you can travel as far as you like across the Universe in as little time as you like, *in under light speed*. But don't think you can return to Earth and find your family and friends still alive.

A fascinating, yet baffling, coda to this discussion is to consider what it must be like for a light beam to travel around in space. In fact, this is where we must push the implications of relativity to their logical conclusions: if you were to ride a light beam, any distance you cover, even across the entire Universe, would shrink to zero – which is OK, since time itself also stands still, so you cover no distance in no time at all! This is yet another reason why nothing can attain light speed: it is just too crazy to contemplate. But light seems to do it without any problem – it is just that no light beam has ever told us what it feels like.

We will explore this idea more closely in the next chapter. For now, after this brief interlude, we can return to the Pole in the Barn Paradox – not only to resolve it, but to appreciate why it is even a paradox in the first place.

The Pole in the Barn again

Let us restate the problem, now that we are fully 'up to speed', as it were, with the predictions of relativity theory regarding the contraction of lengths when moving close to the speed of light. Remember, you are now standing inside the barn watching the pole-vaulter run towards you at high speed. You know that, when stationary, the pole was the same length as the barn. But now it is moving it will appear to you to be shorter, so that its entire length can easily fit inside the barn. In fact, there will be a split-second, if you act fast enough, when you could shut both front and rear barn doors and trap the pole inside.

However, we must also look at the situation from the pole-vaulter's perspective. For him, the pole is not moving (in the sense that it is not moving *relative to him*) and it is the barn that is fast approaching. He therefore sees a squashed, foreshortened barn heading towards him. As he runs through it, the front end of the pole will protrude out of the rear door of the barn *before* the back end of the pole has even got through the entrance. So it would be impossible to shut both doors at once – the pole is just too long to fit.

Is this some kind of optical illusion or is it a physically real effect? After all, surely you and the runner cannot both be right – either both barn doors can be closed at the same time or they cannot.

The paradox, as I stated at the beginning of the chapter, is that you are indeed both right. This is precisely what relativity tells us will happen on the basis of the two statements 'fast-moving objects appear shorter' and 'all motion is relative'.

The resolution lies in what we mean by *simultaneous events*. I said that you, inside the barn, could close both doors simultaneously, trapping the pole. Of course, you quickly open the rear door a moment later, before the pole crashes into it. That doesn't matter; the important thing is that both doors were simultaneously shut for a

Figure 5.2. The Pole in the Barn Paradox

(a) When the pole is not moving relative to the barn, both are the same length.

(b) To the person in the barn, the fast-moving pole is now much shorter and can fit inside.

(c) To the runner, it is the barn that appears squashed and so cannot accommodate the whole pole at once.

short time. But here is how the events unfold according to the runner: as he enters the barn, and before the front of the pole has reached the rear door, he sees it close briefly. A moment later, it opens again to let the pole get through unscathed. A short time *after* that, once the back of the pole has entered the barn, the front door closes. So, yes, he says after he has stopped and wandered back to you to compare notes, both doors did indeed close, but *not at the same time* – that would have been impossible given how short the barn was to him.

This phenomenon of the ordering of events looking different to different observers who are moving relative to each other is a further consequence of Einstein's relativity theory. And, like the other strange results we have encountered, it is no mere theoretical prediction. It really does happen. But, just like the slowing down of time or the shortening of lengths, it is not something we encounter in everyday life. The reason for this is simple: we do not tend to travel around at close to light speed. For most of us, the fastest we will ever move is on board an aircraft. A jet has a cruising speed of just under 1,000 kilometres per hour. This is a millionth of the speed of light. Such 'relativistic' effects are very hard to detect when we move about so slowly.

I sense scepticism – and, frankly, I feel hurt that you have not wholeheartedly bought into all this relativity stuff (or maybe I do you a disservice and you are content with my explanations). Allow me nevertheless to play devil's advocate and make the problem more serious. Remember I stated that, when not moving relative to each other, both the pole and the barn were the same length. So, in principle, if there were no such thing as relativistic length contraction, the moving pole would still fit inside the barn (just) for an instant. But what if the pole were, say, twice the length of the barn? Presumably, the same arguments hold, right? So that, to you standing inside the barn, the pole will appear shorter when it is moving and, provided it's moving sufficiently fast, can still be short enough to fit easily inside the barn. If you hadn't appreciated this point earlier then you will now: this shortening of lengths is no optical illusion. It is not just that the pole *appears* shorter – it really *is* shorter for you, and you can close both doors simultaneously. But if this contraction of the pole is real, then does it mean the atoms that make up the pole are squashed together? More to the point, the pole-vaulter is presumably not immune to this squashing and will appear flattened to you as he runs. Does he not feel uncomfortable about this? The answer is no, he does not feel any different at all (a little out of breath maybe, since he is running so fast) and it is you that he sees flattened inside the squashed-up barn, since for him you are the one doing the moving as you travel along with the barn towards him.

So, if he does not feel squashed up and still sees the long pole he is carrying looking exactly the same as it did before he started running, then surely, *surely*, the shortened pole you see is just an illusion.

Let's test this. What if there is no door in the rear of the barn but a solid brick wall instead? We will not worry about the safety of the runner here; if you have bought into the idea that he can run at near light speed then you can also believe he can safely come to an abrupt halt before he reaches the wall.

Again, we consider how the events play out according to the two points of view. For you, the front door of the barn can indeed still be closed once the shortened pole completely enters the barn, which is *before* the front of the pole hits the brick wall.

But in the runner's frame of reference, the front of the pole crashes into the wall *before* the back of the pole has even entered the barn. If we assume that both the pole and the brick wall are strong enough to withstand the collision and survive intact, how then does the back end of the pole ever even get *inside* the barn for the door to be closed on it? Now, it seems we are faced with a more serious problem than just the ordering of events. It is as though an event – the closing of the door after the pole has entered – does not even take place according to the runner. Surely, we have finally forced a real paradox, one that paints Einstein and his theory into a corner.

Well, no. There is a perfectly valid and correct explanation. You see, in the runner's frame of reference, the front of the pole does indeed hit the wall, but the back of the pole is oblivious to this event, for according to relativity there simply cannot be any such thing as a truly rigid body. Remember I said that nothing can travel faster than light, so the front of the pole cannot transmit the information that it has come to an abrupt halt (some kind of shock wave that it sends down its length) fast enough to stop the back of the pole moving at the same speed it was moving before. Basically, the back of the pole is unaware that the front of the pole has come to a sudden halt. It continues to move so fast that by the time the information reaches it that the front of the pole has stopped, it has already entered the barn, and the door can be closed.

Note here that we must be prepared to open the door again very quickly because the pole won't remain contained inside for long. As soon as the runner has come to his abrupt stop inside the barn, he and you will start seeing lengths for what they really are (by which I mean what they are when not in motion – this is referred to in relativity theory as their 'proper length'). And recall I said that the pole was, in this example, twice the length of the barn. The runner will now agree with you that once the different parts of the pole come to a stop – remember, true rigidity is not possible – they will grow back to their proper length. The front of the pole can go nowhere since it is blocked by the brick wall, and so it is the rear of the pole that quickly expands back through the now opened front door of the barn until half of it is protruding again.

There is one final subtlety here that I will not discuss in detail, but which deserves a mention. In all that I have described, I talk about you and the runner seeing the pole do different things at different times. But even to see the front of the pole, or the rear of the pole, takes time:

this is the time it takes for light to travel to your, and the runner's, eyes. Since the pole is itself travelling at near light speed, such considerations become important. But I will spare you further technical detail for now. Suffice it to say that, as far as our original question is concerned, our piece of string (or pole) varies in length depending on how fast it is travelling. And you can console yourself further in the knowledge that what comes next regarding paradoxes of time builds on the foundations laid down in this chapter. So we will be able to cut to the chase more quickly.

SIX

The Paradox of the Twins

By travelling very fast we can move into the future

IN THIS CHAPTER we continue with the theme of paradoxes arising from the predictions of Einstein's relativity theory. Here, however, we will be delving into the pleasingly mind-bending concepts associated with the nature of time itself and how it is affected by travel at close to the speed of light.

The storyline of this paradox may sound like science fiction, but is in fact perfectly within the mainstream science taught to every physics student as an example of the implications of relativity, even if it is not technologically achievable just yet. It involves a spacecraft capable of reaching near light speed – which, while we have no means of developing such a craft at the moment, is nevertheless perfectly admissible in principle. And, because it doesn't exceed the speed of light, there is no need to appeal to the rather more speculative ideas one finds in popular science fiction such as warp drives or short cuts through hyperspace.

Meet our heroes, the twins Alice and Bob, who design and build this spacecraft. While Bob remains behind on Earth, Alice pilots the spacecraft, flying it from the Earth on a round trip through the galaxy that takes her exactly one year to complete. On her return to Earth she is biologically one year older, she feels that one year has gone by, and all the clocks and timekeeping devices on board the spacecraft agree that a single year has elapsed since she left the Earth.

Bob, meanwhile, has been monitoring her journey throughout, and he witnesses one of the weird effects that manifest themselves as a result of her travelling at near light speed, and is predicted by Einstein's theory of relativity: time on board the spacecraft appears to Bob to be running more slowly than it is on Earth. If he were to watch events inside the spacecraft through a camera he would see everything taking place in slow motion: clocks on board ticking more slowly, Alice moving and speaking more slowly, and so on. So a journey that seems to Alice on board the spacecraft to have taken just one year could in fact last ten Earth years as far as Bob is concerned. Indeed, Alice arrives back on Earth to find that her twin has aged ten years, whereas she is biologically just one year older.

This in itself is not the source of the 'paradox' of the chapter title. It may sound odd, but it is perfectly consistent with the predictions of Einstein's theory. The times I have chosen are arbitrary and depend on just how fast the spacecraft has been travelling. Had Alice, for instance, nudged even closer to the speed of light, then quite straightforward maths for anyone with a pocket calculator (and a little knowledge of relativity) would show that one year on the rocket could correspond to a million years on Earth or, equivalently, during a day's travel by Alice's reckoning thousands of years would go by on Earth. But let us stick to the spacecraft speed that leads to the one-year/ten-years mismatch to at least allow Alice to return while her brother is still alive.

The paradox comes about because of seemingly conflicting conclusions arising from the idea of relative motion. To that extent, our story is rather similar to the Pole in the Barn Paradox of the last chapter; however, messing with the nature of time is always more tricky to get our heads round than issues to do with distances. You see, it appears we have been too hasty in our arbitrary choice of reference frames and the way time slows down in one and not the other. In the last chapter we met the first postulate of relativity: that all motion is relative. Let's see if that applies here. Surely, Alice has just as much right to claim that it is not her spacecraft that is moving away from Earth at near light speed, but the Earth that is moving away from the spacecraft in the opposite direction. What, after all, is true motion anyway? Can Alice not claim that she has been stationary throughout her one-year journey and all the while the Earth has been moving, first away from and then towards her? That this is indeed the case is borne out by the fact that through her camera she would have seen the receding Earth clocks ticking more slowly than the clocks on board the spacecraft! It is therefore Bob, she could claim, who will have aged less on her return, since during her one year away, only a tenth of that time (just over a month) should have gone by on Earth. This is our paradox.

This apparent symmetry in the effects of relative motion has been the source of much confusion over the years. Indeed, many scientific papers have been published claiming to show that this paradox disproves Einstein's theory and its suggestion that time really runs more slowly in one frame of reference than another. Surely, what Bob and Alice see is some kind of optical illusion and time is not *really* slowing down at all. At first glance it might seem that, given the apparent symmetry between the two frames of reference, there should be no difference between the elapsed times on Earth and on the space-ship, so that Alice and Bob remain the same age when she gets back. So are both twins wrong? They certainly cannot both be right.

Believe it or not, the correct answer is that Bob is right. Alice really will have aged less than he has when she returns. The puzzle is how this gets round the business of relative motion. Why is the seemingly symmetrical picture wrong?

To resolve this paradox, I will first need to convince you that time really does slow down close to the speed of light in the same way that lengths alter when objects travel at such very high speeds. First, it is useful to think more carefully about the nature of time itself. This will stand us in good stead for the next chapter, when things really hot up as we encounter the first true paradox arising from time travel.

What is time?

It is fair to say that no one really understands, at a fundamental level, what time is. Our current best theory of time is that provided by Einstein's General Theory of Relativity, which I introduced in the chapter on Olbers' paradox. But even our most powerful scientific theories often fall short when we try to push them too far in trying to answer such deep metaphysical questions as: 'Does time really flow, or is that just an illusion?' and: 'Is there an absolute rate to the flow of time, or even an unambiguous direction to this flow?' Clearly, statements such as 'time points from the past to the future' or 'time goes by at a rate of one second every second' are somewhat unhelpful.

Until Isaac Newton completed his work on the laws of motion more than three centuries ago in his *Principia mathematica*, time was considered to be the domain of philosophy rather than science. Newton described how objects move and behave under the influence of forces, and since all movement and change require the notion of time to make sense, time had to be included as an integral part of his mathematical description of nature. However, Newtonian time is absolute and relentless. It is said to flow at a constant rate as though there were an imaginary cosmic clock that marks off the seconds, hours, days and years regardless of our (often subjective) experience of its passage, and we have no influence over its rate of flow. This all sounds perfectly reasonable; but modern physics has shown beyond question that this view of time is wrong.

In 1905, Einstein announced his discovery that time and space are interconnected in a profound way, publishing his theory of relativity and bringing about a revolution in physics. He showed that time is no longer absolute and independent of the observer. Rather, it can be stretched and squeezed depending on how fast you are moving.

I should make clear at this point that any such variations in the rate of the passage of time have nothing to do with our own, subjective, awareness of it. Of course, on an anecdotal level, we are all familiar with having been at an enjoyable party when the evening has whizzed by far too quickly or, conversely, sitting through a boring presentation or speech that seems to take an eternity to finish. We know that time is not really speeding up or slowing down in these situations. Likewise, we find as we get older that time seems to rush by at an increasing rate. Again, we know that this is not because time itself is speeding up, but because each year that goes by is a tinier fraction of our lives; just think back to when you were a child, and how long it would seem from one birthday to the next. Despite such experiences, we all have a strong gut feeling that some absolute Newtonian time really exists 'out there' and flows at the same rate everywhere in the Universe.

But even before Einstein came along, some scientists and philosophers were unhappy with this view of an external, absolute time, and many had debated issues surrounding the rate of flow, and direction of flow, of time. Some philosophers argued that time itself is an illusion. Consider the following mini-paradox, which sounds rather like something Zeno himself might have come up with:

Time, I am sure you would agree, can be divided up into three components: the past, the present and the future. Although we have records of the past and remember events that have taken place, it cannot be considered to exist any longer. The future, on the other hand, has yet to take place and therefore can also be said not to exist. That leaves the present moment, which is defined as the dividing line between past and future; surely 'now' exists. But

although we 'feel' that the 'now' is a changing moment that is
steadily sweeping through time transforming the future into the
past, it is nonetheless just an instant and as such does not have any
duration itself. The constantly changing present moment, therefore,
is no more than a dividing line between past and future and also
cannot be said to have a real existence. If all three components of
time do not exist, then time itself is an illusion!

You may, as I do, take such clever philosophical arguments with a
pinch of salt.

But back to our main subject: what is much more difficult to justify
is the notion that time actually 'flows'. It is of course hard for us
to deny the feeling that this is indeed what happens, but in science
having a 'gut' feeling about something, however strong that feeling
is, is not enough. In our everyday language we say that 'time passes',
'the time will arrive', 'the moment has gone' and so on. But if you
think about it, all motion and change must, by definition, be judged
against time. This is how we define change. When we wish to describe
the rate of a certain process we count either the number of events
in a unit of time, such as the number of heartbeats per minute, or the
amount of change in unit time, such as how much weight a baby has
put on in one month. But it becomes nonsensical to try to measure the
rate at which time itself changes, since we cannot measure time against
itself.

To clarify this, let me ask you the following question: how would
we know if time were to suddenly speed up? Since we exist within time
and measure the duration of intervals of time using clocks that, like
our internal biological clocks, must presumably speed up also, we
would never be aware of it. The only way to talk about the flow of
(our) time is to judge it against some external, more fundamental,
time.

But if an external time against which we could measure the rate of flow of our own time did exist, then we would only be pushing the problem further back rather than resolving it. Surely, if time by its nature flows, then why should this external time not flow also? In which case we are back to the problem of needing a further, even more fundamental, timescale against which to measure the rate of flow of external time, and so on in a never-ending regression.

Just because we are unable to talk about a *rate* of flow of time does not mean that time does not flow at all. Or maybe time is standing still while we (our consciousness) are moving along it: we are moving towards the future rather than the future coming towards us. When you look out of the window of a moving train and observe fields rushing by you 'know' that they are standing still and that it is the train that is moving. Likewise, we have the strong subjective impression that the present moment (what we call 'now') and an event in our future (say next Christmas) move closer together. The time interval separating the two moments shrinks. Whether we say that next Christmas is moving closer to us or that we are moving closer to next Christmas, it amounts to the same thing: we feel that something is changing. Surely we can all agree on that? I'm afraid not. Many physicists argue that even this idea is not valid.

Strange though it may sound, the laws of physics say nothing about the flow of time. They tell us how everything, from atoms to clocks to rockets to stars, behaves when subjected to a force at any given moment in time, and they provide us with the rules for computing an object's behaviour, or state, at any future time. Nowhere, however, do the laws of physics contain a hint of flowing time. The notion that time *passes* or moves in some way is completely missing in physics. We find that, like space, time simply exists; it just *is*. Maybe this feeling we have that time flows is just that: a feeling, however real it might seem to us. For now, science is unable to provide a satisfactory explanation for where this strong sense we have of passing time and a changing present moment comes from. Some physicists and philosophers even go as far as to say that there is some-thing missing in the laws of physics. They may turn out to be right.

OK, I think we have had enough philosophy for now. Let's get back to understanding how and why the rate at which time goes by changes according to the Special Theory of Relativity, because without sorting this out we will be unable to resolve the Paradox of the Twins.

Slowing down time

Let us, then, look at the nature of time according to Einstein. In the previous chapter I described how lengths are measured to be different by two observers moving at high speed relative to each other. A quick way of seeing how time must also be affected is as follows. A familiar formula everyone learns at school is that speed equals distance divided by time. Now, we know that all observers, no matter how fast they are travelling with respect to each other, see light moving at the same speed. If they measure distances to be different (as they do in the 'pole in the barn' example), then it follows that their time measurements must also be different, such that when they divide their respective measures of distance by time they both arrive at the same (correct) answer for the speed of light. Thus, if one observer measures the distance between two points at 1 billion kilometres, and measures a light beam to cover this distance in one hour, whereas a second observer claims that the two points are 2 billion kilometres apart (remember from the last chapter that no two observers in relative motion agree on lengths), then the second observer would have to measure that light beam taking twice as long to cover the distance, if he is to arrive at the same value for the speed of the light beam. In numbers, the first observer would claim that the light was travelling at 1 billion kilometres per hour; the second observer sees it cover 2 billion kilometres in two hours, or 1 billion kilometres per hour – the same as for the first observer.

Thus, the requirement that we all measure the same light speed forces upon us the notion that time intervals between two events – in this case the start and finish time of the light beam's journey between the two points – will have different values for different observers: an hour elapsing for me may be two hours for you.

Because of the difficulty we all have in grasping this concept of different rates of time, I will make another attempt at convincing you. Imagine that you shine a beam of light from a torch out into the sky, and I then fly up in a rocket alongside the beam, travelling away from you at three-quarters the speed of light. Indeed, you measure the light (moving at 1 billion kilometres per hour) to be pulling away from me at a quarter of the speed at which it is leaving the torch (in the same way that a fast car overtakes a slower one at a speed that is the difference between their two speeds). What would you logically expect me to see if I look out of the rocket window? The obvious commonsense answer is that, like you, I should see the light overtaking me at just a quarter of the speed at which it is travelling away from you. However, since Einstein has insisted that all observers measure light to travel at the same speed, I should in fact see the beam overtaking me at 1 billion kilometres per hour – the same speed at which you see it leaving the torch. This is indeed what the theory of relativity predicts, and this result has been verified thousands of times in laboratories over the past century. But what are the implications?

(A note of caution here: I am using the word 'see' when talking about measuring the speed of a light beam. But of course, for us to see something, light has to travel from it to our eyes. That takes time. And what do we mean anyway when we say we 'see a light beam'? Do we mean bouncing light off the light? So I'm using the term 'see' here just to mean 'measure' in some way – for example, in the case of a pulse of light, by recording the precise time it triggers devices along its path.)

How, then, can I, while travelling alongside the beam of light at three-quarters of its speed according to you back on Earth, still see it moving past me at the same speed at which it leaves the torch? The only way this could be possible is if my time is running at a slower rate than yours. Imagine we both had identical clocks. You would see mine ticking more slowly than yours. Not only that; everything on board my rocket is running more slowly – I even move around in slow motion, and when I communicate with you, you hear that my speech is slower and has a deeper timbre. And yet, I would not feel any different and would be unaware of any slowing down of time.

Students who study Einstein's theory learn how to calculate mathematically the degree to which time will be slowed down given a certain rocket speed. In fact, time on board a rocket travelling at three-quarters light speed relative to some observer will be running 50 per cent more slowly than the observer's clock. That is, every minute that the observer sees ticking by on the rocket clock will take ninety seconds according to the observer's own clock.

You might think that such a situation is of only hypothetical interest, since we do not have rockets that can achieve anything like these speeds. But even at the much more modest speed of, say, the Apollo Moon mission spacecrafts (about 40,000 kilometres per hour), the effect on time is still there, with the travelling clocks and mission control clocks falling out of sync by nanoseconds every second – tiny differences too small to need to be taken into account, but certainly measurable. We will return briefly to this example in a little while.

But let's quickly look at another real-world example, where this effect is important (we will come on to one more a bit later). The slowing down of time in high-speed travel is known as 'time dilation' and is routinely taken into account in physics experiments, particularly those in which subatomic particles are accelerated in 'atom smashers' such as the Large Hadron Collider at CERN in Geneva. There, particles can reach speeds so close to that of light that if such 'relativistic' effects were not taken into account, the experiments would not make any sense.

So, the lesson we learn from Einstein's Special Theory of Relativity is that a consequence of the constancy of the speed of light is that time runs more slowly during high-speed motion. It is at this point that I need to drop another bombshell about time. Recall from Chapter 3 that Einstein in fact produced two theories of relativity: the Special Theory in 1905, and then the General Theory in 1915. It was in the latter that he revised Newton's ideas about the nature of gravity and instead described this force much more fundamentally in terms of the effect of mass on the very fabric of its surrounding space and time.

Einstein's General Theory of Relativity thus gives us an alternative means of slowing time down: through gravity.

The Earth's gravity causes time to run more slowly than it does out in empty space away from the pull of any stars or planets. And since all objects have mass, they will all be surrounded by their own gravitational fields. The more massive the object, the stronger the gravitational pull it exerts on nearby bodies and, according to Einstein, the bigger its effect on time itself. A fascinating consequence of this when applied to the rate that time flows on Earth is that the higher in altitude we go, the weaker the Earth's gravitational pull, and hence the faster time is able to run. In practice, the effect is tiny: we would have to travel very far out into space before we were entirely free of the Earth's gravity. Even at an altitude of 400 kilometres, which is a typical orbit for satellites, the pull of gravity is still 90 per cent of what it would be on the surface of the Earth. (Note that the reason satellites are able to orbit indefinitely without plummeting to the ground is because they are precisely that, in orbit: they are in free fall around the Earth and are therefore weightless because they keep moving.)

An amusing example I like to give when describing gravity's effect on time is to say that if my wristwatch is running slow, then one way to correct it is to hold my arm above my head. Since the watch is now higher up it will feel a slightly weaker gravitational pull and run a little faster. This effect is real, but so tiny as to make the exercise rather pointless. For instance, to make up just one second I would have to keep my arm aloft for several hundred million years!

In some situations, the two types of time dilation effect (due respectively to special and general relativity) can act against each other. Consider two clocks, one on the ground and one in a satellite in

Figure 6.1. Speeding up time

Does the clock on board a satellite run faster or slower than one on Earth? We need to understand both of Einstein's two theories of relativity to work it out.

orbit around the Earth. Which one will be running more slowly? To the clock on the ground, the high-speed motion of the clock in orbit should be making it run slower, while the fact that it is in free fall around the Earth and therefore not feeling any gravity should be making it run faster. Which effect wins?

This is all starting to sound quite paradoxical in its own right, and yet the combination of effects was confirmed beautifully in a remarkable experiment carried out in the early 1970s. It is known today as the Hafele–Keating experiment after the two American physicists who conducted it.

In October 1971, Joseph Hafele and Richard Keating placed very accurate clocks on board two commercial airliners and flew them around the world: one eastward, in the direction of the Earth's rotation, and the second westward, against the rotation, and then compared them against a clock on the ground at the United States Naval Research Observatory in Washington DC.

The two effects on time here, the slowing down of rapidly moving clocks and the speeding up of high-altitude clocks, had to be carefully measured, taking into account whether the planes were flying with or against the Earth's rotation. Let us consider this carefully. Because of their roughly similar altitude, the clocks on board both planes would have felt weaker gravity; this would speed them up relative to the clock on the ground. However, because the eastbound plane was moving with the Earth's rotation and so would be travelling faster (like rowing downstream), its clock would be running more slowly than the one on the ground – whereas the clock on board the aircraft travelling westward, against the Earth's rotation (like rowing upstream), would run a little faster than the clock on the ground.

At the start of the experiment all the clocks were carefully synchronized. At the end, the eastbound clock was found to be 0.04 microseconds (millionths of a second) slow (the slowing down of the clock due to its high speed of travel winning over the opposing speeding-up effect due to weaker gravity at altitude), while the west-bound clock ended up being almost ten times as much (0.3 microseconds) too *fast* (the speeding-up of time due to weaker gravity now enhanced by the effects of special relativity).

It's all rather confusing, and even the smartest physicists have to furrow their brows to get their heads round it, but the important point is that, in both cases, what was measured in the experiment agreed beautifully with the mathematically predicted results from Einstein's theories.

Today, these effects on time are routinely taken into account on board the GPS satellites that map locations all over the surface of the Earth (this is the other real-world example I promised you). Without these corrections for the tiny differences between the rates of flow of time on board the satellites and on the ground, we would not be able to fix our positions on our smartphones and car satnavs to the accuracy we have become used to. This precision in position – to within a few metres – depends on the time it takes for a signal from the device on the ground to bounce off a satellite and come back again, which needs to be accurately timed to within just a few hundredths of a microsecond. So how bad would things get if we ignored relativity? Well, relative motion means that the satellites' clocks will be running slower than ours by around seven microseconds per day. However, the absence of a gravitational pull on the satellites (remember, they are in free-fall orbit) means their clocks will be running faster than Earth clocks by about 45 microseconds per day. This works out at a net speeding up of 38 microseconds each day. Since each microsecond translates to a distance of about 300 metres, ignoring Einstein would mean the satellite getting our position wrong by over 10 kilometres each day – and the effect is cumulative.

Now that I have introduced the idea of gravity slowing time down as well as high speeds quickening it up, let us briefly revisit the example of clocks on board the Apollo Moon missions. This will help us when considering the problem of the twins.

Apollo 8 was the second manned mission in the American Apollo space programme and the first human spaceflight to leave Earth orbit. The three-man crew of Frank Borman, James Lovell and William Anders became the first humans to go far enough away to see planet Earth in its entirety, and were also the first humans to directly see the far side of the Moon. On their return, Frank Borman pointed out that all three astronauts were older than they would have been had they not flown to the Moon. What's more, he joked that they should be paid overtime for the extra fraction of a second they were away compared with the elapsed time on Earth. Although financially insignificant, the additional time on board the mission was very real.

This would seem at odds with the paradox at the centre of this chapter, in which the travelling twin, Alice, comes back younger than her stay-at-home brother. In fact, the reason the effect was the opposite way round is precisely because of the subtle interplay between the two relativistic effects on time. Overall, the three astronauts aged about 300 microseconds more than they would have done had they stayed on Earth. Let's look at how this worked.

Whether time on board Apollo 8 runs faster or slower than on Earth depends on how far away the spacecraft is. For the first few thousand kilometres of its outbound journey, the Earth's gravity is not yet weak enough for time on board to speed up much, and Apollo's speed relative to the Earth remains the predominant factor; with this causing time to run more slowly for the astronauts, they aged more slowly than people on Earth. But as they travelled further from the Earth, the effects of its gravitational pull lessened and Apollo's time began to speed up – implying that the effects of general relativity were winning over those of special relativity. Over the entire journey, this speeding up of time on board was the larger effect and so more time elapsed on the spacecraft than on Earth – hence the 300 extra microseconds.

As a bit of fun, physicists at NASA very carefully checked whether Borman was right about the overtime, and discovered that it was so for only one of the three astronauts, William Anders, who was making his maiden space flight on Apollo 8. Both Borman and Lovell had already completed an earlier two-week orbital mission on Gemini 7, during which, it was calculated, the slowing down of time due to their speed was the dominant effect, and they had therefore aged less than people on Earth by about 400 microseconds. So overall they both achieved a net gain of 100 microseconds and were slightly younger than they would have been had they stayed on the ground. Thus, rather than earning overtime, they had in fact been overpaid!

Resolving the Paradox of the Twins

Now that we have established the effects of gravity on time, let us return to examine and, hopefully, resolve the paradox set up at the beginning of the chapter with Alice and Bob. Recall that each can argue that it is the other who is really moving, so each argues that the other's time has run more slowly: Bob says that Alice has flown off in her spacecraft and returned having aged less than him, whereas Alice argues that it is Bob and the Earth that have gone away and come back again, and so it is Bob whose time has been running more slowly and he who has aged less.

There are several ways of analysing this problem and I have great fun with my students in the course I teach at Surrey University going over the different arguments. First, let us examine the simplest one.

The true answer, as I said earlier, is that Bob is right and Alice is wrong: she will indeed return younger than her brother. First of all, we should note that their situations are not entirely symmetrical. Alice has to speed up as she leaves the Earth and then, if she is travelling in a straight line, must decelerate, turn around and accelerate again, before finally slowing down when she reaches the Earth. Bob, on the other hand, has remained at the same constant speed throughout. Even if Alice follows a circular path so she can continue at the same speed, she will still feel the effects of acceleration as she constantly changes direction. So the twins' relative motion is not entirely symmetrical: Alice feels the effects of her journey, while Bob remains stuck on the gently rotating Earth. However, this does not obviously provide us with the reason for her ageing less.

There is a way of looking at the problem without any speeding up or slowing down. Alice starts out in space and gets up to speed before she passes the Earth, at which point she and Bob synchronize their clocks. She travels in a straight line at constant speed and then, at some point (I know that this is not realistic, but bear with me), she instantaneously switches direction without a change of speed to head back to Earth. This is what physicists refer to as an 'idealized situation' – not possible in practice but serving as a useful simplification without being wrong. We can now analyse the situation in terms of the distance Alice covers as measured by each of the twins; for the reason Alice ages less can be explained by length contraction.

Let's say Alice's turnaround point is the star Alpha Centauri, which is four light-years from Earth (that is, its light takes four years to reach us, and vice versa). If Alice is travelling at half the speed of light, then Bob will calculate that she takes twice as long as light to cover the distance: eight years, giving a total journey time of sixteen years. However, for Alice, the distance she must cover is contracted due to the relativistic effect of her speed – or rather, the speed at which Alpha Centauri is moving towards her, since she can legitimately claim that her spacecraft is at rest. It is now obvious that, for her, the journey time there and back (when she can claim that it is the Earth that is heading towards her) will be less than that experienced by Bob; if she doesn't need to travel so far, she is not going to take so long.

In reality, of course, Alice is not able to make this instantaneous switch in her direction and of course must decelerate, turn around, then accelerate again. It is here that we really must appeal to the other way of slowing time down, due to general relativity. But where is the gravitational effect now? In the example I used the star Alpha Centauri as the point at which she turns around, but that was not necessary. Alice could have turned around anywhere in empty space and never encountered any gravitational field. There is therefore one final idea of Einstein's that we must consider.

The happiest thought of his life

Have you ever wondered why, when describing the effects of acceleration in a fast car or jet aircraft, we refer to the notion of g-force? We say that a racing driver feels several 'g's when accelerating, braking or going round a corner fast. The 'g' stands for 'gravity' and highlights a very important link between acceleration and the force of gravity. We all know what this feels like. When you're on a plane that is just about to take off, you first hear the engines roar as the pilot boosts them to maximum power, then you are pushed back into your seat as it accelerates down the runway, quickly gathering speed before it lifts up into the air. Try moving your head forward away from the headrest before you've taken off and you will experience a force trying to pull it back. This resistance will feel similar to the pull of your head's weight down on the pillow when you are lying on your bed. In fact, if the aircraft accelerates at 1g it will feel exactly the same. Acceleration mimics the effects of gravity.

Einstein came upon this equivalence a few years before he completed the formulation of his General Theory of Relativity. He called it, rather unimaginatively, the Principle of Equivalence. He would later say that this realization, this 'eureka' moment, was the happiest thought of his life – which, if nothing else, shows the comprehensiveness of his dedication to science. He had been contemplating what happens when objects are in free fall. The feeling of weightlessness that we foolishly endure on a rollercoaster ride as we descend best illustrates this equivalence, for it is at that moment, when we surrender to the Earth's gravitational field, that we stop feeling its pull. It is as though our downward acceleration has cancelled out our experience of gravity.

Einstein went on to show that all the effects of gravity on space and time also show up when an object is accelerating. Indeed, if you were sitting in a chair in a spacecraft that was accelerating through space at 1g, then the feeling you'd have would be indistinguishable from the one you'd have if your chair was tipped on to its back on the

Figure 6.2. Slowing down time

Running round in circles at close to light speed will slow your time down.

ground back on Earth. In both cases you would be feeling the same pull holding you down into the back of the seat. This is a crucial idea, because it implies that, just as a gravitational field will slow down the flow of time, so should acceleration. And it does. If you spend time speeding up and slowing down, this is equivalent to immersing your-self in a gravitational field and will be an effect over and above that of the Earth's gravity itself.

So now we can finally lay the Paradox of the Twins to rest. The reason Alice ages less than Bob is because she is the one undergoing acceleration and deceleration, and her time therefore runs more slowly during those periods, according to the predictions of general relativity, whether she travels in a straight line there and back or not. In fact, the more she changes direction by following a wiggly path through space, the more time she will be undergoing acceleration and deceleration, and the less time elapses for her.

Watching the clocks

We could stop here, I suppose. There is no Paradox of the Twins after all – no Clocks Paradox, as it is sometimes rather less imaginatively called – since their 'journeys' through space–time are not symmetrical. But it is interesting to consider what each of the twins would see if they had a way of sending messages back and forth during the journey.

Alice and Bob can agree to send each other a light signal at regular fixed intervals according to their own timing. What if they sent these flashes once a day at the same time? During Alice's outbound journey, they are moving apart at high speed, so each will receive the other's signal at intervals that are greater than twenty-four hours because of the effects on time predicted by special relativity. But in addition to this, each pulse of light will have further to travel than the one before it and so there will be an additional delay over and above the one due to the slowing down of time. This second effect is the same as the principle behind the Doppler shift (the change in frequency, or pitch, of waves produced by moving sources, whether light or sound).

Then, whenever Alice slows down, speeds up or changes direction, her time is slowed down further and her signals arrive further apart. Finally, what is particularly interesting is what happens on her return journey, because the two effects that were combined together, both delaying the arrival of the pulses of light, on her outbound trip are now in competition: the relative high-speed motion between the twins still means each measures the other's clock to be ticking more slowly than their own, but because the light signals they each send have progressively *shorter* distances to travel as they approach each other, they start to arrive bunched up. The maths shows that this bunching up (pulses arriving more frequently than every twenty-four hours) wins over the effect of the slowing down of time and they would see each other's clocks speed up. In fact, they would observe each other's motions and actions to be running at higher speed too. Nevertheless, the bottom line after all these considerations is that Alice returns to Earth having aged less than Bob.

Is there any more to say on the matter? Well, yes. For here, at last, is the payoff. If Alice has been travelling for one year according to her and returns to an Earth where ten years have elapsed, has she then not just time travelled nine years into the future?

The poor man's time travel

Many would argue that slowing time down is not real time travel. After all, is this really any more impressive than suspended animation or even, come to think of it, sleep? If you drop off and wake up thinking you've been asleep for just a few minutes, and then, on checking your watch, see that several hours have passed, isn't this also a bit like time travel into the future?

I would argue that the relativistic slowing of time is much more impressive and is indeed real time travel, albeit the poor man's version. You might think that true time travel to the future implies that the future must be already out there, existing alongside our present and awaiting our potential arrival. This is not the case here. What is happening is that the future is unfolding on Earth all the time that Alice is away. It is just that, since less time elapses for her, she is moving on a different time track from the Earth's. In a sense, she is fast-forwarding into the future and arriving there before everyone else. How far into the future Alice manages to go depends entirely on her spaceship speed and how winding her flight path is.

The real question, then, is this: if Alice comes back to Earth and doesn't like what she sees, is there a way for her to return to her own time? This, of course, requires time travel back to the past, and is a whole different kettle of fish. In fact, it is an issue that leads on to the one true paradox in this book – one that we will explore in the next chapter.

SEVEN

The Grandfather Paradox

Going back to the past and killing your grandfather means you would never have been born

IF YOU WERE to travel back in time and murder your maternal grandfather before he met your grandmother, your mother would never have been born and neither would you. But if you were never born, your grandfather could not have been killed by you and would instead have lived to meet your grandmother, and so you would have eventually been born, travelled back in time and killed him, and so on. The argument goes round for ever in a self-contradicting circle. It seems you are unable to murder your grandfather because you are there to try.

This is the classic time-travel paradox and it comes in many guises. For instance, it has always puzzled me why you would need to travel back far enough to kill your grandfather rather than your mother or father – maybe skipping a generation makes it less horrible. It does not have to be so brutal, either; this is just how it has been traditionally stated – in more violent times, presumably. For instance, a much more benign version would involve your building a time machine, then travelling back in time and destroying it before the moment that you use it, so you are now unable to travel back and destroy it.

The paradox can also be stated another way. A scientist discovers on a shelf in his lab instructions to build a time machine. He follows them and, one month later, uses the time machine he's built to travel back one month in time, taking the instructions with him. He places them on the shelf in his lab for his younger self to find.

Clearly, as with the grandfather paradox, the future seems predetermined and we no longer have any free choice in our actions. In the first paradox, you cannot kill your grandfather because he must survive any attempt on his life in order to ensure your existence in the first place. In the second example, the scientist must build the time machine because he did/does/will (tenses get a little confusing when discussing time travel). But what if he finds the instructions with a note attached explaining that they were placed there by his time-travelling self from the future, and he then decides against building the time machine and destroys the instructions instead?

There is yet another paradox embedded in this story that is easy to miss, which comes about because the time-machine instructions seem never to have been created in the first place: they were found, used and returned, trapped in a continuous time loop. Where did the information come from? How did the atoms of ink find themselves arranged so carefully on the surface of the paper? It takes intelligence and knowledge to have created those instructions, and yet they seem to be caught in a logically consistent circle, forwards in real time and backwards via the time machine, from which there is no escape and, more importantly, no initial entry point or origin for their creation.

These days, we are all familiar with the notion of time travel back to the past via the many sci-fi stories in books and movies: just think of such blockbusters as *The Terminator* and *Back to the Future*. Most of us are happy to suspend our disbelief – and quite rightly so – so as not to spoil our enjoyment of these stories, but it's easy to get into logical tangles if that is what you are looking for.

Figure 7.1. The paradox of time travel

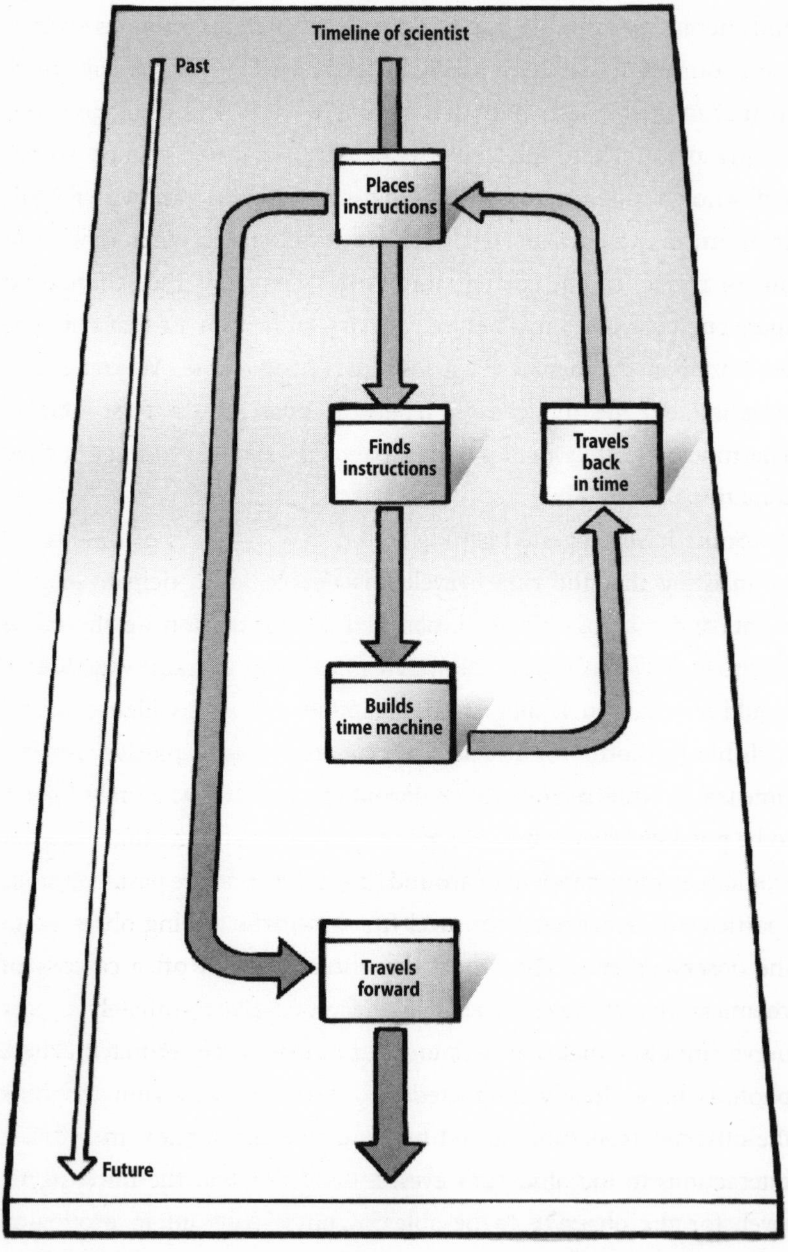

There is a third and final paradox, which we will also need to address: using a time machine violates the law of conservation of mass and energy. For example, you could travel back to five minutes ago and meet yourself so that there will be two of you existing at the same time. At that moment, your body has suddenly appeared out of nowhere, adding extra mass to the Universe. Be careful here; this is not like the well-known phenomenon in subatomic physics known as pair creation, in which a particle and its antimatter partner (a sort of mirror image) can be created out of pure energy. You see, there is no spare energy around just before your arrival in the past that could be used to pay the debt of your sudden appearance. We really are violating one of the central tenets of physics: the First Law of Thermodynamics, which says, to paraphrase, that 'you cannot get something for nothing'.

Some have suggested getting round the paradoxes of time travel by insisting that the time traveller is unable to participate in past events and can act only as an observer. In this version we should be able to go back to the past and watch it unfold in the same way that we would watch a movie, being immersed in the action while remaining invisible to those around us. Unfortunately, such a passive form of time travel, while seeming to be devoid of paradoxes, is even less likely to be possible. That is because to see something – as the time traveller would see events happening around him as he visits the past – photons (particles of light) need to travel from the object being observed to the observer's eyes. They must then be able to set off a cascade of chemical and electrical events in the retina that ultimately trigger nerve impulses that are sent to the brain to be interpreted. These photons have already interacted in a very real way with anything the observer is looking at and have carried information from those interactions to the observer's eyes. Indeed, down at the microscopic level, for the observer to be able to touch, feel and in every way interact with the past he has to be able to exchange photons with his surroundings, because almost any kind of contact between two bodies in the real world takes place, at a fundamental level, via an electromagnetic interaction that involves exchanging photons. I don't want to get too technical, but the bottom line is this: if you see something, then you should also be able to touch it. So, if we can travel back in time to observe the past we should also be able to interact with it and participate fully in events.

If we are to avoid paradoxes arising from our meddling with the past, there will have to be another way of going about it.

How can we reach the past?

There are essentially two ways of travelling to the past. The first is by sending information *backwards* through time. This type of time travel was the inspiration for the science-fiction writer Gregory Benford in his 1980 novel *Timescape*, which features communication between scientists several decades apart: researchers send a message back in time from 1998 to 1962 to warn of impending ecological disaster. They do this using hypothetical subatomic particles called tachyons, the existence of which is predicted by the mathematics of Einstein's relativity, but which have such strange properties that these days they are confined to the pages of sci-fi novels. You see, tachyons (whose name, from the Greek *tachys*, meaning 'fast', was coined in the 1960s when they were studied seriously for a while) are particles that travel faster than light. In doing so they must also travel backwards in time.

This implication was famously and amusingly described by a British-Canadian biologist called Reginald Buller in a limerick he published in *Punch* magazine in 1923:

> There was a young lady named Bright
> Who travelled far faster than light.
> She went out one day
> In a relative way
> And returned the previous night.

We will come back to how and why this is the case later on.

The other way of getting to the past is by travelling what appears to you, the time traveller, to be forwards in time (your own clock runs forwards normally), while moving along a curved path through space–time that takes you back to your past (like looping the loop on a rollercoaster). Such loops are known in physics as 'closed time-like curves' and have been the subject of serious theoretical research in recent years.

Clearly, my mentioning tachyons and time-like curves suggests that I am not about to dismiss time-travel paradoxes out of hand. That would be far too easy: just saying that time travel back to the past (as opposed to the sort of time travel into the future we explored in the last chapter) is logically impossible would make this a short chapter. Instead, we are going to try to resolve these, the toughest scientific paradoxes we have encountered thus far, within the bounds of what the currently understood laws of physics allow. The reason I am taking them seriously is that – and this may come as a surprise to you – it has been known since the middle of the last century that Einstein's theories of relativity actually allow for the possibility of time travel into the past, albeit under certain conditions, and then only because of quirks in the mathematics. His Special Theory of Relativity shows how the first type of time travel (backward causation via faster-than-light travel) is possible, while his General Theory of Relativity permits the existence of the other, more 'traditional' form of time travel via time-like curves. The logician Kurt Gödel, who worked with Einstein at Princeton during the 1940s, demonstrated mathematically that such time travel into the past was at least theoretically possible without violating any laws of nature – apart from the paradoxes we have encountered, that is. We are therefore going to have to confront these paradoxes head-on if we are to rescue Einstein's reputation.

Faster than light

Let us first deal with the issue of why travelling faster than light can mean moving backwards in time. To do this, I will make use of the pole in the barn scenario we met in Chapter 5. To recap, you are standing in the barn watching the runner carrying his length-contracted pole and coming towards you at near light speed. Since, for you, the pole is shorter than the barn, you can close both the front and rear doors simultaneously to trap the pole inside the barn for a fraction of a second. In principle, you could even close the front door as soon as the rear tip of the pole has cleared it but *before* you close the rear door; since the pole is shorter than the barn there will be a brief time between the rear of the pole entering the barn (and the front door closing behind it) and the front of the pole reaching the rear of the barn (by which time that door must be open again to allow the pole through). It is during this very brief window of opportunity that you could close the rear door. So, to reiterate, it is possible in your frame of reference to close the front door of the barn, followed by the rear door.

Now, what if the closing of the rear door is triggered by the earlier closing of the front door? We now have a fixed ordering of events with the rear door only closing (the 'effect') *because* the front one already has (the 'cause'). This necessity for a cause to come before its effect is known as 'causality' and is a crucial concept in nature. Seeing an effect precede its cause violates causality and can lead to all sorts of logical paradoxes. For example, if I flick a switch to turn a light on, then my action is the cause and the illumination of the room the effect. But suppose another observer moving past me at close to the speed of light sees the light come on before I flick the switch. He could then, in principle, stop me from doing so *after* he sees the light come on. For according to the 'relativity of simultaneity', two observers moving relative to each other at near light speed see not only different time intervals between events but sometimes, if events are close enough together in time, even a reversal of their order. This type of backward causation paradox is precisely what happens when signals are allowed to travel faster than light.

To see this more clearly, let us return to the pole in the barn example. The runner, remember, sees a shortened barn that his pole is never going to fit inside. In his frame of reference, which is every bit as valid as yours as you stand stationary inside the barn, the two doors opening and closing must follow a certain order: the rear door must close, and open again to allow the front of the pole through, *before the front door is closed*. Only with this ordering of events can the pole pass through the barn unhindered and yet still allow each door to close briefly at some point. But if the rear door closes only because it is sent a signal from the front door once that has closed, then the runner will be seeing the events taking place the wrong way round, with effect coming before its cause. And we have a problem.

Nevertheless, relativity theory is able to explain this all perfectly well and is backed up by solid mathematics. Consider the following scenario. You set up an experiment such that if you flick a switch on Earth, a light flashes on the Moon. It takes light about 1.3 seconds to cover the distance between the Earth and the Moon, so if your signal to the Moon travels at light speed then you will see the flash through your telescope 2.6 seconds later (the time needed for the 'there and back' journey of light). But what if you could send the signal faster than light? If you see the flash just 2 seconds later it would mean that the time taken between flicking the switch and the flash taking place is just 0.7 seconds (2 minus 1.3). This all sounds perfectly feasible, but relativity theory shows us why it should not be allowed in nature.

To convince yourself of this conclusively, you really have to do the sums – alternatively, you can just take my word for it. To someone in a rocket travelling to the Moon at close to the speed of light, the flash of light on the Moon would be seen to happen *before* you flick the switch on Earth. They could then send you their own faster-than-light signal to say they had seen the Moon flash. This signal would appear to you to be travelling backwards in time and you could even receive it *before* you flick the switch. You might then decide *not* to flick the switch. The only way to avoid such a situation is to rule out faster-than-light signalling.

This is one of the reasons why physicists believe nothing can travel faster than light, since if anything could it would lead to a genuine paradox. This, I would argue, rules out the first type of time travel.

But what about the idea of time-like curves looping around through space–time?

The block universe

In order better to visualize space–time paths I need to introduce the concept of the block universe, which is a simple yet profound way of picturing space and time in a unified way.

Imagine the universe as a vast rectangular box. Now, what if we want to add time as another dimension? This gives us a combined four-dimensional volume of space–time that is referred to as the 'block universe'. However, since we cannot think in four dimensions we have to make an obvious simplification if this idea is to give us a picture that is of any practical use: we sacrifice one of the dimensions of space by squashing down the three dimensions of volume into a two-dimensional sheet that makes up one surface, or side, of the block universe. Then the dimension that heads off from left to right at right angles to this surface (the third dimension) can be used as the time axis. Think of it as a giant loaf of sliced bread, in which each slice is a snapshot of the whole of space at a single moment, while successive slices correspond to successive times. Of course, this is not really accurate since space is three-dimensional not two-dimensional, but it does give us a helpful way of visualizing the time axis. Figure 7.2 shows you how this works to help us visualize the block universe.

What is neat about this picture is that any event happening at a certain place and at a certain time can be represented by a point in the box (**x** in Figure 7.2). More importantly, we now see the whole of time

Figure 7.2. The block universe

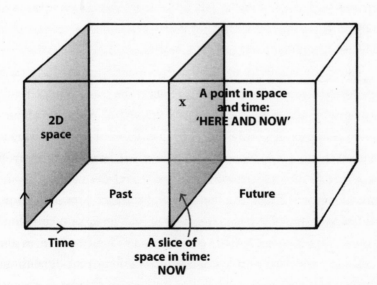

laid out before us – a timescape, as opposed to a landscape – with all events, whether in our past or future, coexisting in this timeless, static block universe.

But does this have anything to do with reality, or is it nothing more than a useful visualization tool? For instance, how do we make the connection between this static model of space–time and the real sense that time 'flows'? Physicists have two ways of viewing it. Common sense tells us that our 'now' is a slice of space, with the Universe at earlier times represented by the region to the left of this slice and the Universe at future times to the right. This view of the totality of existence, in which the whole of time – past, present and future – is laid out frozen before us is one we can never have in practice since we cannot extract ourselves from the Universe. Our 'now' moves along from left to right, jumping from instant to instant, from one bread slice to the next, like frames in a movie.

The alternative is to do away with any notion of a present moment altogether, so that past, present and future coexist and all events that have ever happened or ever will happen sit alongside each other in the block universe. In this picture, not only is the future pre-ordained but it is already out there and as unalterably fixed as the past.

This is, in fact, far more than a convenient visualization. It is the view forced upon us by the way space and time are interwoven in the real Universe as described by Einstein's theory of relativity. Think of two separate events, A and B, which may or may not be causally linked, where A takes place before, and in a different location from, B. According to our understanding of space and time before Einstein came along, the spatial and the temporal distance between events A and B were assumed to be independent of each other and the same for all observers. However, Einstein has shown how two observers moving relative to each other and measuring these two quantities (time and distance) will not come up with the same values for either. But stick them together within space–time and we see that, within the block universe, all observers agree on a single 'distance' between the two events that is a combination of a space part and a time part. Only in space–time do we get some absolute number we can all agree on. This turns out to be crucial in relativity. Of course, that is not what is interesting to us here – I just felt I should mention that we don't make up stuff like the block universe just for the hell of it.

The coexistence of all times in the block universe makes the notion of time travel seem far more possible. If we are able to travel back in time to a particular moment, then as far as the people there are concerned we will have arrived in their present moment, their 'now', having arrived from the future. To them, the future is just as real as the present. And in any case, what makes our 'now' any more special than theirs? We cannot claim that our present is the true 'now' and that they just think they live in the present, since we can similarly imagine time travellers from our future visiting us in our present; to them, we are the past. So, both our future and our past – indeed, all times – must exist together and are all equally real. This is precisely what the block universe model tells us.

Time travel in the block universe

Fundamentally, no one really knows how time 'flows', if indeed it does at all. But at least we can assign to it a direction, an arrow of time. This is an abstract concept that means we can define an ordering of events. An arrow of time points from the past to the future, from earlier events to later ones. It is a direction in time in which things happen. Think of it as forced upon us by the Second Law of Thermodynamics. It's like the arrow on the 'Play' button on your DVD player: even though you are free to fast-forward or rewind back and forth as much as you like, still the movie runs in a particular direction, not the other way.

Despite this constraint, the block universe is beginning to look like one vast DVD movie in which we have the freedom to jump from one time to another, in the past or the future. There is no true present moment since every point in the movie is just as real as any other; they all coexist. Is it, then, possible to control time in the real universe in this way? Are all past and future times really 'out there', somewhen, being acted out, and thus no less real than what we perceive as our now? If so, how do we reach them? This is the crucial question. We know we can move from one point in space to any other, so why not through time also?

A possible solution to time-travel paradoxes

When faced with difficulties in testing the predictions of their theories, physicists sometimes resort to what are called 'thought experiments' – idealized imaginary scenarios that do not violate any laws of physics and yet are too impractical or hypothetical to set up as real experiments in the lab. One of these is called 'the billiard table time machine'. With it, we can ask what would happen if something were to travel back in time and meet itself. What would the mathematics predict?

The idea is that a ball drops into the pocket of a billiard table. The pocket is linked via a time machine to another of the pockets with a spring mechanism, such that the ball is spat back out on to the table at an earlier time than when it dropped into the original hole. This allows for the possibility of its colliding with itself before it went in.

In this thought experiment certain paradoxes can actually be easily avoided if we allow only those situations that do not lead to a paradox in the first place. Physicists refer to these as 'consistent solutions'. Thus, a ball can go back in time, pop out of another pocket and deflect the earlier version of itself such that it goes into the hole, enabling it to travel back in time in the first place. However, the situation in which the ball emerges from the pocket and collides with its earlier self such that it causes it to miss the pocket would not be allowed since this leads to a paradox.

The underlying idea in all time-travel paradoxes is this: in our universe there is just one version of the past. It has already happened and cannot be altered. In principle, we can travel to the past and meddle with history as much as we like, provided that whatever we do only causes things to turn out the way they have. We can never alter the course of history because we are an integral part of the Universe and carry with us memories of how events unfolded in our past. Basically, what has happened has happened.

We can even think up scenarios in which it is only *because* the time traveller went back and meddled with the past that things turned out the way they did, just as in the billiard-table time machine.

So does this insistence on allowing only 'consistent solutions' in time travel help to resolve *all* the paradoxes? The answer is a resounding 'no'. Superficially, one can see its appeal: you can travel back in time to meet your younger self only if you remember a time in your past when you were visited by your older time-travelling self. If you do not, then the meeting never did, nor ever will, take place. Likewise, in the more violent Grandfather Paradox scenario, you cannot kill your grandfather since the proof of your failure (for whatever reason) is that you were born to become the time traveller.

However, this does not help us get round the other paradoxes I have mentioned, such as the time machine instructions that are caught in a time loop without being created. (The only way out in that example is if the younger version of the scientist finds the instructions, destroys them without studying them, then works out how to build a time machine and creates identical new instructions that he can take back and place on the shelf. You see, it is not enough for the scientist to destroy the instructions after he looks at them, for then the information on how to build the time machine is itself caught in the time loop.)

Finally, the consistent solutions argument still does not explain the violation of the First Law of Thermodynamics when the time machine and its contents arrive unannounced in the past, adding to the mass and energy of the Universe at that moment in time, even though they have been 'borrowed' from the future.

True time travel needs a multiverse

By now we have dealt with most theories of time travel, but let's take a look at one of the most bizarre, yet remarkable, ideas to emerge from theoretical physics in the past half-century: that of parallel universes. It was devised in its original form to explain away some of the weirder implications and observations of the quantum world, whereby atoms can be in more than one place at once and are able to behave both as tiny localized particles and spread-out waves depending on how we decide to study them, and where two particles can seemingly communicate instantaneously with each other even if they are on opposite sides of the Universe. Such phenomena may seem like paradoxes in their own right, and we will come back to them in Chapter 9 when we get to meet Schrödinger's cat. But what is particularly interesting about this theory of parallel universes right now is what it has to say about the possibilities for time travel.

In its earliest incarnation, the idea of parallel universes was known as the 'many worlds interpretation' of quantum mechanics, according to which, as soon as a subatomic particle is faced with a choice of two or more options, the whole Universe splits into a number of parallel realities equal to the number of options available to the particle. There are, according to this view, an infinite number of universes, which differ from the one we inhabit to a greater or lesser extent depending on how long ago they split from ours, and each of these universes is just as real as our own. It seems like a crazy idea at first sight, but when set alongside some of the equally crazy implications of quantum physics it turns out to be no less credible.

For several decades the many worlds interpretation remained a curiosity in physics and the stuff of science fiction. To date, no experimental evidence whatsoever has been found that parallel universes actually exist, and we have no way of being able to make contact with any of the others. It might seem impossible that there could be room for all these other worlds and dimensions. After all, it may well be that our own universe is itself infinite in extent. Where can all the others be? The way to think about them is that they are like overlapping block universes. They all share the same single time axis, but each one has its own spatial dimensions, coexisting on top of each other but never interacting other than at the quantum level.

More recently, the many worlds idea of universes branching off from each other has been replaced with a more sophisticated theory called the Quantum Multiverse. According to this idea, the Universe does not divide up into multiple copies of itself all the time; instead, there are already an infinite number of parallel universes coexisting and overlapping, each just as real as the others. Suddenly, our block universe has just got very crowded. But there are advantages to this idea over the single block universe in which there was just one future, fixed and static. Now, all possible futures are open again and we can reclaim our free will. The choices we make lead us along pathways through all possible space–times – and it is our selection of pathways that ultimately defines our universe. The infinite selection of possible futures open to us represents the infinite totality of universes coexisting in the Multiverse.

Suddenly, true time travel becomes possible, for our space–time contains just one of an infinite number of futures *and* an infinite number of pasts. Travelling back into the past in the Multiverse is no different from the way we would normally get carried along into the future – just as there are many futures available to us, so there are many pasts that we could visit. Time travel would involve following a time loop into one of these possible pasts. This means that a time-like curve into the past will almost inevitably slide us into the past of a nearby parallel universe. Think of it this way: if you could have your time again and tried to make the same choices by following the same actions, then, no matter how hard you tried, something would be different second time round – not necessarily because you made a slightly different choice, but probably because something else, somewhere, has followed a different path through space–time, altering the future – and you will end up in a slightly different future. The same would happen if you went back to the past: you never end up in the past of your own universe, since that is exceedingly improbable. What is overwhelmingly likely is that you will slide into the past of a universe almost identical to your own. In fact, given the complexity of any one universe it would be almost impossible for you to be able to tell the new one apart from the original – until, that is, you started to meddle.

Once there, you are free to alter the past as you wish, since it is no longer your own past. Events in the parallel universe you have travelled to need not turn out the way they did in your own universe. One thing to remember, though, is that the chances of being able to find your way back home to your own universe are very small – there are simply too many to choose from.

Let us now see how the multiverse theory resolves the Grandfather Paradox and other time-travel paradoxes. We'll begin with the original one. You are now able to kill your grandfather (although this is still not a very nice thing to do) in the new universe you have arrived at. All that happens is that you will never be born *in that universe.*

The example of the scientist and the time machine instructions is also clarified. The time-travelling scientist slides into a parallel universe in which his younger version can choose whether or not to make use of the instructions to build a time machine. No paradoxes arise since he will never become the time-travelling scientist.

Even the problem of non-conservation of mass and energy is resolved, because it now applies no longer to each universe separately, but to the whole Multiverse. The energy and mass which you are made of has moved from one universe to another, so the total sum of mass and energy in the whole Multiverse hasn't changed.

Linking the universes

One of the problems we have yet to address in the multiverse picture is the thorny issue of causality. It seems as though the parallel universe you end up travelling to must know in advance that you will be arriving there. Because the end of the journey (now in the new universe) taken by the time machine is at an earlier moment than the start in your universe, not only does your unannounced arrival have to satisfy the laws of physics in that universe, but the choices you make and the changes you bring about would not have happened had you not travelled back. Is this really an improvement on the paradox of travelling back to your own universe's past? It seems that events having already happened in a parallel universe force you in the future of your own universe to travel back in time. Are we allowed to violate causality if cause and effect are in different realities altogether? It still sounds dodgy, doesn't it?

There is a way out of this, but it requires the time machine to be built and switched on, not at your end but at the end you arrive at in the past. The link between the universes is then made *forward in time*. Once in place, this link should allow two-way travel between the universes. And general relativity permits a way, at least theoretically, of connecting up our universe with a parallel one in just this way. It is known as a space–time wormhole.

Wormholes are hypothetical constructs in the fabric of space–time itself that are not thought to exist in the real world, but since they are admissible in theory (and in our current best theory of the nature of space and time at that) then we can at the very least entertain the possibility that they *could* exist. Unlike their cousins, black holes, which most physicists and astronomers are now confident really exist out in space and are formed when matter gets squeezed down under incredible pressure, such as in a collapsed star or at the centre of galaxies, wormholes can only be formed under very special conditions that we do not think exist naturally in our universe. Nevertheless, in theory at least, a wormhole would be a shortcut through space–time, out of our universe, either linking back to it in a completely different place and time or joining it with a parallel

Figure 7.3. A wormhole in space–time

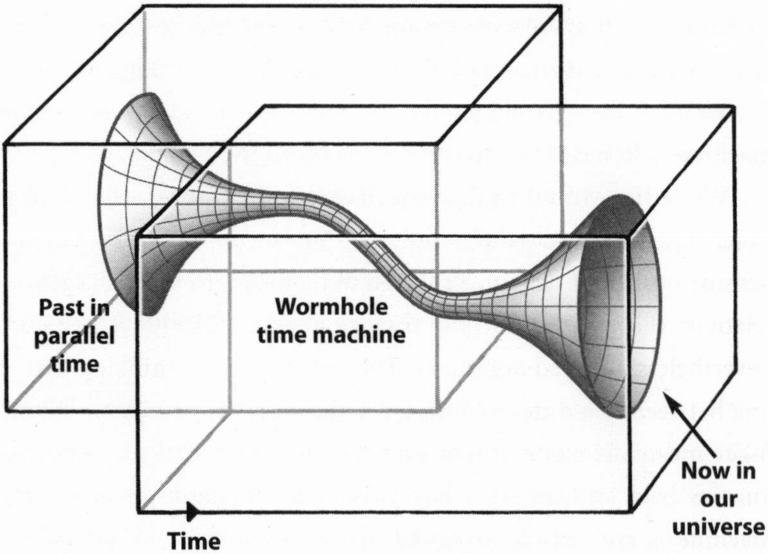

To avoid paradoxes, a wormhole would need to be linked up to the past of a parallel universe.

universe. It is such space–time tunnels that finally provide us with the hope of time travel.

Have we then satisfactorily relegated the Grandfather Paradox and its relations to the status of merely 'perceived paradoxes' that have melted away under the onslaught of physics? Actually, no; while I have highlighted *possible* ways of resolving the paradoxes, in doing so I have drifted into the realm of speculation. I have not broken any laws of physics, of course, but ideas like the multiverse and wormholes in space–time remain just outside conventional science: fun to consider, but impossible to verify . . . for the time being, anyway.

Where are all the time travellers?

Many people have used this question as their argument against the possibility of time travel. If, they claim, time travel into the past is realized, then surely there will be some time travellers who have chosen to visit our time and they should be walking around among us. But, so far, we have never encountered any. Surely this is proof that time machines will never be built?

While time travel to the past may indeed turn out to be impossible, either because parallel universes or wormholes do not exist or because of some as yet undiscovered refinement of Einstein's theory of relativity that rules it out, the absence of time travellers among us is nevertheless a flawed argument. The mistake lies in thinking that the link between two different times, whether via a wormhole or by some other means, is made at the moment the time travellers begin their journey back in time. It is not. It is made at the moment the time machine is created (or switched on) allowing for the *possibility* of time travel. If we work out how to build a time machine in the twenty-second century, then it can be used after the moment it is activated to travel back to as early as that activation time, but never as far back as the twenty-first century. This is because constructing a time machine involves linking different times together within the Multiverse. All times before this would have been lost for ever so that they are no longer 'available'. This rules out any possibility of our ever being able to go back to prehistoric times – unless we stumble upon a naturally occurring time machine, say a very old wormhole somewhere in space–time.

So one reason there are no time travellers among us today is simple: time machines just haven't been invented yet.

There are in fact a whole host of other reasons for the absence of time travellers. For instance, if the multiverse theory is correct – and I would argue it has to be if time travel is to be allowed – then our universe is simply not one of the lucky ones visited by time travellers (assuming time machines have already been invented in a parallel universe). Another reason could be that time travel to the past is forbidden by some as yet undiscovered laws of physics. Or there may be a more mundane reason: expecting to see time travellers among us presupposes that they would, in fact, want to visit this century. Maybe for them there will be much nicer and safer periods to visit. Or it just might be that time travellers from the future are indeed among us, but choose to keep a low profile.

EIGHT

The Paradox of Laplace's Demon

Can the flapping of a butterfly's wings rescue us from a predictable future?

'PREDICTION IS VERY difficult, especially about the future.' So said the Danish quantum physicist Niels Bohr. The quote might sound trite and frivolous, but hidden behind it, as was so often the case with the utterances of Bohr, are profound ideas about the nature of fate, free will and our ability to determine how the future will unfold.

Let me first set up the paradox. The French mathematician Pierre-Simon Laplace devised his own imaginary demon half a century before Maxwell proposed his. Laplace's demon is far more powerful than Maxwell's since it has the ability to know the exact position and state of motion not merely of every air molecule in a box, but of every particle in the Universe, and it fully understands the laws of physics that describe how they interact with each other. This means that, in principle, such an all-knowing demon could work out how the Universe will evolve over time and be able to predict its state in the future. But if so, it could choose to act on this information and deliberately carry out an action that would cause the future to evolve differently from its earlier prediction, thus rendering its prediction wrong and undermining its own ability to foresee the future (since surely it would have taken into account in its computation the action it would carry out).

Here is a fun example that brings the paradox into sharp focus. Imagine the demon is in fact an enormous supercomputer that is so powerful and has so much memory capacity that it can know every tiny detail about the Universe, right down to the state of all the atoms that make up the computer itself and every electron flowing through its circuitry. With this information it is able to compute precisely how the future will unfold. It is given the simple instruction by its operators, which it presumably predicted it would receive, to self-destruct if it computes a future in which it still exists, but to do nothing if it computes a future in which it no longer exists (since it will have self-destructed). To repeat: if it predicts a future in which it is still around, then it won't be, and if it predicts a future in which it is not then it will be. Either way, its prediction is wrong. So, will it survive or not?

Like many of the other paradoxes in this book, the resolution of this one tells us something profound about reality and strays far beyond mere philosophical debate. Laplace himself seems to have been unaware of the paradoxical nature of his demon; indeed, he referred to it simply as 'an intellect'. Here is his original description:

We may regard the present state of the Universe as the effect of its past and the cause of its future. Consider an intellect that, at a certain moment, would know all forces that set nature in motion, and all positions of all items of which nature is composed. If this intellect were also vast enough to submit these data to analysis, it would embrace in a single formula the movements of the greatest bodies of the Universe and those of the tiniest atom; for such an intellect nothing would be uncertain and the future, just like the past, would be present before its eyes.*

Laplace was not looking for a paradox – he was using this hypothesis to highlight something widely believed to be incontestable at the time: that the Universe is 'deterministic'. This word is central to the paradox of this chapter, so we need to understand what it means and define it carefully. Determinism means that the future can, in principle, be predicted. However, the above paradox suggests that we must rule out this possibility: that Laplace was wrong and our universe cannot be deterministic. But, as we will see, subject to certain caveats and uncertainties in our current physical theories, we have every reason to believe that our universe is indeed deterministic.

Does this imply that we must discard our ideas of free will, since our fate is already sealed? And how then do we resolve the paradox of Laplace's demon?

We can briefly compare this situation with the time-travel para-doxes of the last chapter. In that case, our past was fixed and known to us, but we had to travel to it in order to change it and force paradox. Here, Laplace's demon knows the future, but doesn't require time travel; it just waits for the future to come to it and, while waiting, it can meddle with the present to force a different future to evolve.

One not very scientific way of ruling out time-travel paradoxes is to insist that time travel to the past is simply impossible. But in the case of Laplace's demon, no time travel is necessary; the demon cannot escape the future, which is heading its way even if it does nothing, so it looks like we need another explanation to resolve the paradox. The simplest option is often the correct one, and here it must surely be that, unlike the past, which is fixed, the future is still open and yet to

*Pierre-Simon Laplace, *A Philosophical Essay on Probabilities* (1814), trans. F. W. Truscott and F. L. Emory, 6th edn (New York, Dover, 1951), p. 4.

be determined. What the demon can presumably 'see' is just one possible future. Surely, in order for it – and, hopefully, us – to be able to make free choices, our universe cannot be deterministic. It's an appealing argument, but not necessarily the one needed to resolve the Paradox of Laplace's Demon.

To show you why this simple resolution is not sufficient, imagine the following scenario. You use the supercomputer to calculate the state of the Universe in the future – a future in which we have a wonderful new theory of physics that we have hit upon after decades of ingenious experiments and advances made with contributions from many great scientific thinkers. This theory is encapsulated in a set of beautiful mathematical equations. You get the computer to tell you this information, thereby relieving you of the need to have gone through the long process of scientific research that leads to it in the future. You thank the computer, switch it off and go off to claim your Nobel Prize without doing any of the work.

Here is the problem: if the computer really had predicted one of a potentially infinite number of possible futures, which just happened to be the one in which this profound scientific discovery is made, then we can see that there is no element of real prediction at all here: it is no different from its having hit upon the idea through sheer chance. It is somewhat similar to the popular 'infinite monkey theorem' in which a monkey hitting keys randomly on a typewriter for an infinite length of time will at some point, by sheer accident, type out the complete works of Shakespeare. We have therefore learned nothing from this explanation. And while it is not impossible that the computer could come up with a new scientific 'theory of everything' by accident like this, that outcome is so exceedingly improbable that we can ignore it. Admittedly, of course, the computer will have started its calculation from the present moment and taken into account the current state of knowledge and the trains of thought of the world's leading theoretical physicists, as well as the ideas for new experiments that could be carried out in the future, so it is somewhat less improbable than a monkey tapping away randomly at a keyboard coming up with the same theory – but the chance of that outcome emerging is still vanishingly small.

However, there is of course a perfectly valid way out of the paradox, and I might as well come clean with it now. If I sound a bit reluctant, it is because this resolution is rather more trivial than the paradox deserves. I mentioned when describing the power of the supercomputer that its knowledge is so complete it even knows all the details about its own inner structure and hence could predict its own actions (forget the issue of whether or not the computer has free will, since we will assume that, despite its immense power, it is not self-aware and does not know it can trick itself by doing something other than what it predicts it will do). Where the scenario comes unstuck is when we analyse what it means for the computer to know the state of every atom, every electron it is made of. It needs to store this information in its memory banks, which are themselves made of atoms that are arranged in a very special way, which is itself part of the very information it holds, and so on – which is of course paradoxical, and rules out the possibility that the computer can know all about itself. It cannot therefore include itself in the calculations it makes in predicting the future, and this means that its knowledge of the Universe is incomplete.

The above argument is sufficient to rule out Laplace's demon. But is this, then, all we need to say about the paradox? Not at all. In highlighting the possibility of knowing the future we have opened up a Pandora's box of issues relating to whether or not we live in a deterministic universe, what this has to say about our own freedom to carry out actions, and whether the future is pre-ordained and set. Science has something to say about all these issues.

Determinism

Let me begin by carefully distinguishing between three concepts: determinism, predictability and randomness.

First, by 'determinism' I mean what philosophers refer to as causal determinism: the idea that events in the past cause events in the future. It therefore follows, taking the idea to its logical conclusion, that everything happens as the result of a chain of events that can be traced all the way back to the birth of the Universe itself.

In the seventeenth century Isaac Newton came up with his laws of mechanics using the newly understood mathematics of calculus, which he was instrumental in developing. His equations allowed scientists to predict how objects will move and interact with each other, from the firing of cannonballs to the motion of the planets. Using his mathematical formulas, values for the physical attributes of an object, such as its mass, shape and position, along with its speed and the forces acting on it, could be plugged into simple equations that could provide information on the state of the object at any future time.

This led to the widely held belief, which lasted for the next two centuries, that if all the laws of nature could be known it would in principle be possible to compute the future action of every object in the Universe. Ours was a universe in which everything – all movement, all change – was predetermined. There was no free choice, no uncertainty and no chance. This model became known as the Newtonian clockwork universe. At first glance, it is not as bleak as Einstein's block universe, in which everything that has ever happened and will ever happen in the future is laid out frozen in time before us. But in fact, the clockwork universe is no different in the sense that its state at all future times is predetermined and fixed.

Then this view suddenly changed. In 1886 the King of Sweden

offered a prize of 2,500 kroner (a tidy sum – more than most would have earned in a year) to whoever could prove (or disprove) the stability of the solar system: that is, say for sure whether the planets would continue to orbit around the Sun for ever or if there was a chance that one or more of them might one day spiral into the Sun or escape the pull of its gravity and float away. The French mathematician Henri Poincaré took up the challenge. He began by looking at a simpler problem involving just the Sun, the Earth and the Moon – what is referred to as a three-body problem. He discovered that even with just three bodies, the problem was mathematically impossible to solve exactly. What's more, certain arrangements of the three bodies would be so sensitive to initial conditions that the equations pointed to completely irregular and unpredictable behaviour. He won the King's prize even though he didn't come up with an answer to the original question about the stability of the whole solar system.

Poincaré had discovered that the way a system of even just three interacting bodies evolves in time cannot be knowable exactly – let alone one involving all bodies in the solar system (at least, all the planets and their moons, along with the Sun). But the implications of this discovery were not explored for another three-quarters of a century.

The butterfly effect

Let's give our all-powerful computer the far more modest task of predicting the way the balls on a pool table will scatter when struck by the cue ball at the start of a game. Every ball on the table will be knocked in some way and most will undergo multiple collisions, many bouncing off each other and the sides of the table. Of course, the computer would need to know precisely how hard the cue ball is struck and the exact angle at which it collides with the first ball in the pack. But is this enough? When all the balls have finally settled, how close might the computer's prediction of the distribution be to reality? While it is perfectly feasible in theory to predict the outcome when only two balls are involved in a collision, it is almost impossible to take into account how the complicated multiple scattering of many balls will end up. If one of the balls moves at a slightly different angle it might collide with another ball, which it might have originally missed, so that both trajectories are dramatically altered. Suddenly, the final outcome looks very different.

So it seems we must feed into the computer not only the initial conditions of the cue ball, but the precise positions of all the other balls on the table: whether they are touching each other, what the precise distances between them and the cushions are, and so on. And even this is not enough. A minute speck of dust on any one of the balls would be enough to perturb its path by a fraction of a millimetre, or cause it to slow down by a minuscule amount, slightly altering the force with which it hits another ball. We also need to give the computer accurate information about the precise state of the table's surface: where it is slightly more dusty or worn, for example, so that its friction with the balls is slightly more or less.

Still, you can imagine that the task is not impossible. It can, *in principle*, be done provided we have all the information about the initial conditions and fully understand the laws and equations of motion. The way the balls end up is not random – they all obey the laws of physics and behave in accordance with the forces acting on them at any given moment in time in a perfectly deterministic way. The problem is that we can never make a completely reliable prediction about this *in practice*, because we would need to know all the initial conditions to an incredible degree of precision, which includes taking into account every particle of dust on every ball, and every strand of fibre in the cloth of the table. Of course, if there is no friction between the balls and the table then they will continue to collide and scatter for much longer, and we would therefore need to know the initial positions of the balls to an even greater degree of precision to determine where they will end up when they do finally stop.

This inability to know, or control, the initial conditions, along with all other continuing influences, to an infinite degree of accuracy can be found even in seemingly much simpler systems. For example, when tossing a coin it is simply too much to expect to be able to repeat the same action and achieve the same outcome time and again. If I toss a coin and get 'heads', it is too difficult for me to toss it in exactly the same way again, making it spin the same number of times so as to land 'heads' again.

In both the pool-table example and the tossing of a coin we could, if we had complete knowledge, repeat exactly the same action to achieve an identical final outcome. This repeatability is the essence of Newton's world and is found everywhere. But so is the sensitivity to initial conditions. And we see it everywhere in daily life. If you make a certain decision one morning on your way to work, like pausing for a second before crossing the road, then you might miss the opportunity of bumping into an old friend who gives you some information that leads to your applying for a new job that changes your life; a split second later still in crossing that road and you could be hit by a bus. Our destiny may be mapped out for us in a deterministic universe, but it is totally unpredictable.

The man who first brought these ideas to the world and in doing so helped create the new concept of chaos was Edward Lorenz, an American mathematician and meteorologist who hit upon the phenomenon by accident while he was working on modelling weather patterns in the early 1960s. He was using an early 'desktop' computer, the LGP-30, to run his simulation. At one point he wanted to repeat a simulation by running the computer program again with identical inputs. To do this he used a number the computer had calculated and printed out halfway through its run. He typed it back in and set the program running again. He assumed the computer would arrive at the same outcome it had the first time. After all, the number it was using was the same, wasn't it?

Actually, no. The computer had the ability to calculate to a precision of six decimal places, but the printout it produced rounded this up to just three decimal places. It had used the number 0.506127 in its original run, but this was printed as 0.506, and that is what Lorenz fed in the second time. He had assumed that this tiny difference between the two values (0.000127) would lead to only a tiny difference at the end of the simulation, however long it kept running. It didn't. And that was a huge surprise. Lorenz had discovered that tiny changes can sometimes produce very big effects. The simulation was an example of what we now understand as nonlinear behaviour. This is why it is so difficult to make long-term weather predictions, since we can never know to infinite accuracy all the variables that affect the real weather. It's just like the pool-table example, only far more complicated. We can now know with reasonable reliability if it will rain in a few days' time, but we can never know if will rain on this date next year.

It was this profound realization that led Lorenz to coin the term 'the butterfly effect'. The idea of the flap of a butterfly's wings having a far-reaching ripple-type effect on subsequent events seems to have first appeared in a short story called 'A Sound of Thunder', written in 1952 by Ray Bradbury. The idea was borrowed by Lorenz, who popularized it as the now familiar notion of the flapping of a butterfly's wings somewhere leading months later to a hurricane on the other side of the world. Of course, it is important to clarify that this does not imply that the hurricane will have had its origins in the flapping of one particular butterfly's wings, but rather that it is the result of the cumulative effect of trillions of tiny disturbances in the atmosphere all over the world – had *any one of them* been different or absent, the hurricane might not have happened at all.

Chaos

The word 'chaos' in general language is taken to mean formless disorder and randomness, in the sense that a children's birthday party can be chaotic. In science, the term 'chaos' has a more specific meaning. It mixes determinism and probability together in a way that is not at all obvious. Once understood, it can be perfectly logical and intuitive, but the fact that this understanding was gained relatively recently shows just how unexpected it was. Here is one definition of chaotic behaviour: if a system behaves in a cyclical manner, repeating the same actions over and over again, but the way it evolves depends sensitively on its initial conditions, then it won't pass through exactly the same state on each cycle but will *appear* to be behaving randomly, changing its path in a totally unpredictable way.

Chaos is not really a theory as such (although 'chaos theory' has become a commonly used term, and I still plan to use it). It is a concept, or phenomenon, that we find to be almost ubiquitous in nature and has spawned a whole new discipline in science with the rather less imaginative title of *nonlinear dynamics* – a description which derives from the main mathematical property of chaotic systems, namely that cause and effect are not related in a linear, proportional way. By this I mean that it had been assumed before chaos was fully understood that, while effect must follow cause, simple causes would always lead to simple effects and complex causes to complex effects. The notion that a simple cause could lead to a complex effect was quite unexpected. This is what mathematicians mean by 'nonlinear'.

Chaos theory tells us that order and determinism can breed what appears to be randomness. In fact, it says that our universe can still be deterministic, obeying fundamental physical laws, while very often showing a tendency to become highly complex, disordered and, most importantly, unpredictable. We now find chaos in almost every area of science. It may have begun in our attempts to understand the weather, but we now find it in the motion of stars in galaxies, in the orbits of planets and comets in our solar system, in the way animal populations grow and decline, in the way metabolism works inside cells and in the beating of our hearts. It is found in the behaviour of subatomic particles, in the working of machines, in the turbulent flow of liquids through pipes and of electrons through electrical circuits. However, it can be most easily seen mathematically, via computer simulations. Although modelling chaotic behaviour mathematically is straightforward, since it involves repeating a simple mathematical formula over and over again, it often requires fairly high computational speed to carry out these simple steps a very large number of times.

To summarize, then, what chaos theory has shown us is that, leaving aside (for the moment) quantum randomness, as far as we know our universe is completely deterministic but not predictable. Yet this unpredictability is not the result of any true randomness: the deterministic nature of the Universe means that it follows perfect and well-defined rules, some of which we have uncovered, others of which we have possibly yet to discover. The unpredictability, then, arises from the impossibility of our ever being able to know, to an infinite degree of accuracy, the initial conditions for the evolution of anything beyond the simplest of systems. There will always be a tiny error in our inputs to the calculation, and this will always have an ever-growing ripple effect that leads to a wrong prediction.

There is also a fascinating and possibly even more important flip side of chaos: that the repeated application of the same simple rules which lead to chaotic behaviour, starting from organized regular motion, can sometimes lead from bland, structureless form to the emergence of beautiful and complex patterns – that we can get order and complexity where there was none before. You start with something without structure, allow it to evolve, and you begin to see structure and patterns spontaneously emerging. This idea has led to the spawning of new academic disciplines known as emergence and complexity theory, which are beginning to play a major role in many diverse areas, from biology to economics to artificial intelligence.

Free will

When it comes to what all this has to say about the nature of free will (and therefore about the Paradox of Laplace's Demon), there are still many different philosophical views and the issue is far from resolved. All I can do is give you my opinion as a theoretical physicist. You are free to disagree with me. Or are you?

There are four options available when it comes to the sort of universe we live in:

1   Determinism is true, so all our actions are predictable and we have no free will, just the illusion that we are making free choices.

2   Determinism is true, but we can still have free will.

3   Determinism is false; there is built-in randomness to the Universe, allowing us free will.

4   Determinism is false, but we still don't have free will since events happen randomly and we have no more control over them than we would if they were predetermined.

Scientists, philosophers and theologians have debated whether or not we have free will for thousands of years. I'm going to focus here on certain aspects of the nature of free will and its connection with physics. I certainly won't be straying into the realms of what is called the mind–body problem, the nature of consciousness or the human soul.

Our physical brains, each consisting of a network of a hundred billion or so neurons linked together via hundreds of trillions of synaptic connections, are, according to everything we know about them so far, nothing more than sophisticated and hugely complicated machines that run the equivalent of computer software, albeit involving a complexity and interconnectedness far beyond anything a modern computer can achieve. All those neurons consist ultimately of atoms that obey the same laws of physics as the rest of the Universe. So if we could, in principle, know the position of each atom in our brains and what it was doing at any given moment, and if we understood fully the rules that govern how they all interact and fit together, then we should *in principle* be able to know the state of our brains at any time in the future. That is, with enough information I could predict what you will do or think next – provided, of course, you are not interacting with the outside world, otherwise I would need to know everything about that too.

Were it not, therefore, for the weird and probabilistic quantum rules according to which those atoms behave, and in the absence of any non-physical, spiritual or supernatural dimension to our consciousness – for which we have no evidence – we would have to admit that we too are part of Newton's clockwork, deterministic universe and that all our actions are pre-ordained and fixed in advance. In essence, we would have no free will.

So do we have free will or don't we? The answer, I believe, despite what I have said about determinism, is yes, we still do. And it is rescued not by quantum mechanics, as some physicists argue, but by chaos theory. For it doesn't matter that we live in a deterministic universe in which the future is, in principle, fixed. That future would be knowable *only if* we were able to view the whole of space and time from the outside. But for us, and our consciousnesses, embedded *within* space–time, that future is *never* knowable to us. It is that very unpredictability that gives us an open future. The choices we make are, to us, real choices, and because of the butterfly effect, tiny changes brought about by our different decisions can lead to very different outcomes, and hence different futures.

So, thanks to chaos theory, our future is never knowable to us. You might prefer to say that the future is pre-ordained and that our free will is just an illusion – but the point remains that our actions still determine which of the infinite number of possible futures is the one that gets played out.

Consider the situation not from the point of view of an individual looking out at the deterministic yet unpredictable world around him or her, but by examining the complexity of our brains and how they work. It is precisely this unavoidable unpredictability about how a complex system such as our brain works, with all the thought processes, memories and interconnected networks with their loops and feedbacks, that gives us our free will.

Whether we call it true freedom or just an illusion in a way does not matter. I can never predict what you might do or say next if you really want to trick me because I cannot in practice ever model every neuronal activity in your brain, anticipate every changing synaptic connection and replicate the wing-flapping of every one of those trillions of butterflies that constitute your conscious mind as I would have to do in order to compute your thoughts. That is what gives you free will. And this is so despite the fact that the actions of the brain are most probably fully deterministic – unless quantum mechanics has a bigger say in the matter than we currently understand.

The quantum world – randomness at last?

Quantum mechanics, the theory of the subatomic world, describes nature's rules down at the tiniest scales, where things are fundamentally different from those in our everyday world. It was realized in the early part of the twentieth century that we cannot use Newtonian mechanics to describe the way a microscopic particle, such as an electron, moves.

If the electron is over here now, and we apply a certain force to it by, for instance, switching on an electric field, then we should be able to say with certainty, and to some degree of accuracy, where it will be one second later. But it turns out we cannot make such a definite prediction, and the reason for this inability seems to involve much more than simply our inability to know the initial conditions precisely enough. The Newtonian equations of motion that govern the behaviour of everyday objects, from coins to pool balls to planets, are useless in the quantum world, where they are replaced by a new set of rules and mathematical relations. These describe a microscopic reality that does indeed seem to be truly random. Here at last, it would seem, we have found the antidote to the fatalistic fixed determinism of the Universe described by Newton and Einstein, for here is where we see what is known as true 'indeterminism'.

Thus, as we saw back in Chapter 2, an atom might radioactively 'decay' by spitting out an alpha particle. But we are unable to predict when this might happen. According to the standard interpretation of quantum mechanics, this has nothing to do with our ignorance of all the necessary information, as was the case earlier. It turns out that we cannot, even *in principle*, predict when the atom will decay, however accurately we could pin down the initial conditions. In a sense this is because the atom itself does not know when this might happen. This uncertainty seems to be a fundamental feature of nature itself at this level, where things behave in a very 'unpindownable' way.

The radioactive atom is not behaving completely randomly, of course, because we find that where there is a large number of identical atoms there emerges a statistical average to their behaviour. The time it takes for exactly half the atoms in a sample of a particular element to decay radioactively is known as the half-life of that element, and this value is something that can be known very accurately with a large enough sample, in the same way that tossing a coin many times will converge on a fifty–fifty probability of its landing heads or tails. However, whereas for the coin this probabilistic feature of the outcome arises from the unpredictability of initial conditions affecting a deterministic process, for the atoms quantum probabilities seem to be built into nature itself, and we can never, even in principle, do any better.

The important question here, then, is whether this quantum indeterminism rescues us from the bleak determinism of our macroworld and gives us back real free will. Some philosophers think it does. They are wrong, in my humble view, and there are two reasons why I make this claim. First, it has been found in recent years that quantum fuzziness and randomness leak away very quickly as we build up complex systems involving trillions of atoms. By the time we scale up to the Newtonian world of people and brains, quantum weirdness has averaged out and evaporated and normal determinism is restored.

The second argument is one that I am personally quite partial to and which cannot be dismissed out of hand. It is just possible that quantum mechanics is not the whole story, and that the unpredictability of a process such as radioactive decay is indeed due to our ignorance. What we may be lacking is a deeper understanding of nature whereby we would, after all, be able to predict exactly when any given atom might decay, in principle if not in practice, just as a fuller knowledge of all the forces involved in the toss of a coin would allow us to predict its outcome. If this is the case we may need to go beyond quantum mechanics to find the answer, or at least develop a different interpretation of the quantum rules. Einstein himself held this view, and this is what he meant by his famous remark that 'God does not play dice'. Einstein could not quite accept the randomness of the quantum world.

Although Einstein's version of the argument has since been shown to be wrong, there is another way of interpreting quantum theory that does not conflict with the standard version, but according to which the subatomic world behaves completely deterministically. It originates with the work of a physicist by the name of David Bohm and has been known for over half a century. The problem is that no one has been able to find a way to check whether this version of quantum theory is right and so categorically confirm or deny whether the Universe is indeed deterministic all the way down to subatomic scales.

According to Bohm, the quantum world's unpredictability arises not from true randomness but from the existence of information that will always remain hidden from us and without which we cannot make precise predictions. The quantum world is unpredictable not because we can never burrow down deep enough into it, nor because of a quantum 'butterfly effect' and sensitivity to the precision of our measurements, but rather because we are simply unable to probe the subatomic world without disturbing it in some way. By looking to see what an electron is doing we unavoidably change its behaviour, rendering our prediction useless. It's a little like asking you to remove a coin from the bottom of a glass of water with your fingers, but without getting them wet. In David Bohm's version of quantum theory, every particle in the Universe has a quantum force field that controls its actions; by measuring the particle, we disrupt this field and so change the particle's behaviour. We still do not know if this description of the quantum world is correct or not, and we may never do.

A final summing up

We have come a long way from Laplace's demon. While the paradox I set out at the beginning of the chapter turned out to be relatively easy to resolve, it did lead on to some fascinating questions about the nature of fate and free will. It would seem that we can never know the future, not because it is random, but simply because it is unpredictable, despite its following regular well-determined rules. This unpredictability is, for some, enough to give us some semblance of free choice. Quantum mechanics, although constrained to defining the very smallest scales, may possibly give us back true randomness, but even this is still in dispute.

And when it comes to the workings of the human brain, no one can be sure when the next breakthrough will come. It might even turn out that the probabilistic nature of the quantum world does indeed have a direct impact on the world of the very large, particularly inside living cells, and possibly the brain. We may have resolved the Paradox of Laplace's Demon; but in doing so we have not answered all these questions.

NINE

The Paradox of Schrödinger's Cat

*The cat in the box is both dead and alive –
until we look*

IN 1935 ONE of the founders of quantum mechanics, the Austrian genius Erwin Schrödinger, had had enough of the weird inter-pretations of its mathematics. Following lengthy discussions with, among others, Albert Einstein himself, he proposed one of the most famous thought experiments in the history of science. He wrote a lengthy paper entitled 'The Present Situation in Quantum Mechanics', which was published in a leading German scientific journal. It has since become known simply as the 'Schrödinger's cat paper', and it is incredible how many people, quantum physicists included, have since tied themselves up in knots trying to emphasize or explain away the supposed paradox Schrödinger described in it. Over the years, many fantastically exotic resolutions of the problem have been proposed, from messages sent backwards in time to the power of the conscious mind to alter reality.

Schrödinger asked what would happen if we were to shut a cat in a box for a while along with a Geiger counter and a tiny amount of radioactive material. So tiny, in fact, is the amount of this material that there is a 50 per cent chance that, over the course of one hour, just one of its atoms might decay, and in so doing spit out a subatomic entity such as an alpha particle. If this happens, it will trigger the Geiger counter and, through a relay mechanism, a hammer will be activated that shatters a small flask of hydrocyanic acid which, upon being released into the box, instantly kills the cat. (Of course, I hope I don't need to explain that no such experiment has ever actually been carried out; this is why it is called a 'thought experiment'.)

As we saw in the last chapter, the moment of decay of a radioactive atom is one of those quantum events that cannot, even in principle, be predicted in advance. According to the standard interpretation of quantum mechanics, which was mostly elucidated and 'clarified' by two of its other founding fathers, Niels Bohr and Werner Heisenberg,

Figure 9.1. Schrödinger's cat

this is not because we don't have all the information we need to make such a prediction, but rather because, at the quantum level, nature itself does not know when it will happen. This is the sort of random event that many believe rescues us from the Newtonian determinism I discussed in the last chapter. All we can say is that there is a certain probability that an atom will have decayed after a certain time (related to the radioactive half-life of the material). At the moment we close the lid of the box we know that no atoms have yet decayed. Thereafter, not only do we not know whether one has decayed or not, we are quite literally forced to describe each atom in the radioactive sample as being in two states at once: both having decayed and not having decayed, with the probability of one rising as the other drops over time. I must stress that this is not just a matter of our ignorance through not being able to see what is going on inside the box. It is forced upon us because this is how the quantum world works: atoms and other entities in the micro-scopic world behave in ways that can only be understood if they are able to exist in this ghostly in-between state. We simply would not have been able to make sense of our world if atoms didn't behave like this.

There are many phenomena in nature that can only be explained if this is indeed what happens. For example, to understand how the Sun shines we must describe the process of thermonuclear fusion taking place inside it in terms of this strange quantum behaviour. The commonsense laws of physics that apply in our everyday macroworld fail to explain how atomic nuclei can fuse together to release the heat and light emitted by the Sun, without which of course we could not exist on Earth. If the nuclei of atoms didn't behave according to the quantum rules, they would never be able to get close enough together to fuse, because their positive electrical charges create a repulsive force-field barrier between them. It is only because they behave like hazy spread-out quantum entities that they can overlap, and so occasionally find themselves on the same side of this force field.

Schrödinger, while acknowledging that the quantum world is indeed very strange, argued that since the cat is also made up of atoms, each single one of which obeys the rules of quantum mechanics, once its fate becomes intertwined (the technical term is 'entangled') with that of the radioactive atom – as it does inside the box – it should also be described according to these same quantum rules. If the atom has not decayed then the cat is alive; if it has, then the cat is dead. Therefore, if the atom is in both states at once, the cat must also exist in two states simultaneously: a live cat and a dead cat. This means that it will be neither truly alive nor truly dead, but in a fuzzy, unphysical, 'in-between' state that only gets resolved one way or the other when we open the box. This is what standard quantum mechanics tells us. And it sounds like nonsense. After all, we never see the cat in this dead-and-alive state when we look, yet quantum physics tells us that is how we must describe the state of the cat *before* we look.

However nonsensical it sounds to you, be in no doubt that this fantastical notion is not just the kind of crazy conclusion reached by theoretical physicists who have spent too long locked away with their equations, but a serious prediction of one of the most powerful and reliable theories in science.

Surely, I hope you will argue, the cat *must* be dead *or* alive, and our opening of the box cannot influence the outcome. Isn't it just a matter of our not knowing what has already happened (or not happened yet)? Well, this was precisely the point that Schrödinger wished to highlight. Despite contributing much to the new theory himself – indeed, the most important equation in quantum physics bears his name – Schrödinger was unhappy about certain aspects of it and had even had several intellectual run-ins with Bohr and Heisenberg during the 1920s over just this issue.

However carefully it is explained to the non-physicist, quantum mechanics will sound utterly baffling, even far-fetched. But the fact is that the rules and equations that describe the behaviour of the quantum world are unambiguous and well defined, both logically and mathematically. And while many quantum physicists themselves do not always feel comfortable with the way the abstract symbols in their equations relate to the real world, the rich mathematical framework of quantum mechanics has been too successful and is too accurate for there to be much doubt that it is reflecting fundamental truths about the world. So is it possible to solve the paradox of the cat, while still retaining quantum mechanics along with all its weirdness? Let's see if we can't solve this puzzle too – after all, we've not come this far, fighting off mighty demons, only to be foiled by a pussycat.

Erwin Schrödinger

The period 1925–7 saw a revolution in science the like of which has not been seen before or since. There have, of course, been other great moments in the history of science, and the advances made by the likes of Copernicus, Galileo, Newton, Darwin, Einstein, Crick and Watson have fundamentally changed our understanding of the world; but I would argue that none of the discoveries of these great geniuses quite revolutionized science to the extent that quantum mechanics did. The field was developed in the space of a few short years and in doing so changed our view of reality for ever.

Let me briefly describe the state of physics in the early 1920s. It was known by then that all matter is ultimately made up of atoms, and scientists had a rough idea of what the interior of these atoms looked like and what they consisted of. It was also known, thanks to the work of Einstein, that light could be made to behave either like a stream of particles or like a spread-out wave, depending on the sort of experiment that was set up and what property of light was being studied. This was strange enough – but evidence was growing that matter particles, such as electrons, could also exhibit such contradictory behaviour.

In 1916 Niels Bohr had returned triumphantly to Copenhagen from Manchester, where he had helped Ernest Rutherford develop a theoretical model of how electrons orbit the nucleus inside atoms. Within a few years he had set up a new institute in Copenhagen, funded by money from the Carlsberg Brewery. Then, with the 1922 Nobel Prize in Physics under his belt, he set about gathering around him some of the greatest scientific geniuses of the age. The most famous of this 'brat pack' was the German physicist Werner Heisenberg. While recovering from a bout of hay fever on the German island of Heligoland during the summer of 1925, Heisenberg made a major advance in formulating the new mathematics needed to describe the world of atoms. But it was a strange kind of mathematics, and what it told us about atoms was even stranger. For instance, Heisenberg argued that not only could we not say exactly where an atomic electron was if we weren't measuring it, the electron itself did not have a definite location but was spread out in a fuzzy, unknowable way.

Heisenberg was forced to conclude that the atomic world was a ghostly, semi-real place that only crystallized into sharply defined existence when we set up a measuring device to probe it – and even then this device would only reveal to us those features it was specifically designed to measure. So, without getting into too much technical detail, while an instrument that measured an electron's position would indeed find it to be in a certain location, and another instrument that could measure how fast the electron was moving would also provide us with a definite answer, it is impossible to set up an experiment that can simultaneously tell us precisely where an electron is and how fast it is moving. This idea was encapsulated in the famous Heisenberg Uncertainty Principle and is still one of the most important concepts in science.

In January 1926, around the same time that Heisenberg was developing these ideas, Erwin Schrödinger submitted a paper outlining an alternative mathematical approach that presented a different picture of the atom. His atomic theory suggested that, rather than the position of an orbiting electron being fuzzy and unknowable, it was in fact like an energy wave surrounding the atomic nucleus. The electron didn't have a fixed location because it really was not a particle but a wave. Schrödinger wanted to make the distinction between a blurred picture of an electron that makes it look fuzzy and smeared out, and a sharply focused picture of an electron as a cloud or bank of fog. In both cases, we cannot specify where the electron is exactly, but Schrödinger preferred to think of the electron as 'really' spread out – until we look, that is. His version of atomic theory became known as 'wave mechanics' and his now famous equation described how these waves evolve and behave over time in a fully deterministic way.

Today, we have learned to live with these two ways of viewing the quantum world: Heisenberg's abstract mathematical way and Schrödinger's wavy way. Both are taught to students and both seem to work fine, with quantum physicists learning to swap easily between the two pictures depending on the problem to hand. The thing is, they both make the same predictions about the world and both agree beautifully with experimental results. Indeed, other quantum pioneers, such as Wolfgang Pauli and Paul Dirac, showed in the late 1920s that the two approaches were completely equivalent mathematically and it was simply a matter of convenience which one was used to describe a particular feature of atoms and their constituents. You can think of this as being a little like describing the same thing in two different languages.

Figure 9.2. Three pictures of the hydrogen atom with its single electron orbiting the nucleus

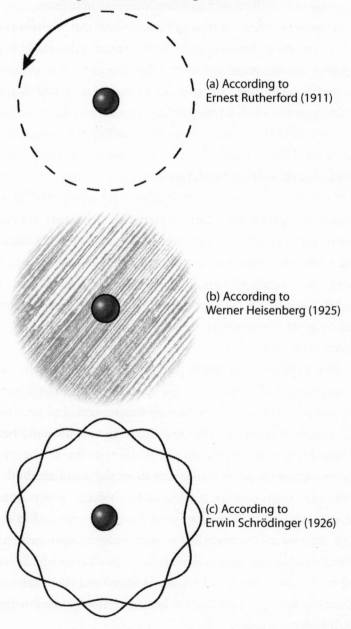

(a) According to
Ernest Rutherford (1911)

(b) According to
Werner Heisenberg (1925)

(c) According to
Erwin Schrödinger (1926)

So, while quantum mechanics as a mathematical theory has been tremendously successful in explaining the structure of the microworld of atoms and all the other building blocks of matter, from electrons to quarks to neutrinos, it is still attended by certain unresolved issues, to do with both how we interpret the mathematics and how the quantum world scales up to the familiar world of large things, the macroworld that we inhabit. It was this second issue that Schrödinger was highlighting in setting up his paradox.

Quantum superposition

There is, though, an important missing step in this story. I realize I am asking you to puzzle over cats, consisting of trillions of atoms, being dead and alive at the same time, while expecting you to buy into the notion that individual atoms can be in two states at once simply because the quantum world is 'weird'. It is therefore probably a good idea to explain what makes physicists so sure that atoms really do behave in this way.

The 'doing two (or more) things at once' or 'being in two (or more) places at once' nature of quantum objects is a property known as *superposition*, and is not quite as unfamiliar as you might think. The idea of superposition is in fact not unique to quantum mechanics at all, but is a general property of waves. We can see this most clearly if we consider water waves. Imagine watching an Olympic diver. As he enters the water you see circular waves travelling outwards from the point of entry all the way to the sides of the pool. This is in stark contrast to the state of the water when a swimming pool is full of people, all splashing about. The water's surface is now turbulent, owing to the combined effect of many disturbances all added together. This process of adding different waves together is called superposition.

The overlap of many waves is rather complicated, and it is easier to consider the superposition of just two waves. Imagine dropping two pebbles into a still pond at the same time, one from your right hand and one from your left. Each one will create circular ripples as it enters the water that spread out and overlap with the ripples created by the other pebble. If you could take a snapshot of this superposition you would observe a complex pattern containing, in certain places, two extremes: there will be regions where two wave crests combine to give a larger swell (called 'constructive interference'), and there will be others where the crest of one wave exactly cancels out the trough of another leaving that spot on the pond's surface temporarily flat as though no wave were passing through it at all (called 'destructive interference'). Hold this idea in your head: the idea that two wave disturbances in a superposition can cancel each other out.

Now let us see the equivalent of this in the quantum world. Devices known as interferometers – which we encountered back in Chapter 5 in the discussion of light and the way its waves travel through space – are able to put two waves together to show the equivalent of where they combine constructively or destructively. An interferometer which would produce some kind of signal we could observe when a single wave enters it could be tuned so that when a second wave enters, it destructively interferes with the first such that the signal disappears. This would be a sure sign that we have wave-like behaviour entering the interferometer.

Now, here is the really cool part. Certain types of interferometer can detect the arrival of subatomic particles, such as electrons. These particles can be sent through a device that splits up their possible path into two, such that they can follow two different possible routes before the paths combine again at the end. If such a device were set up to accept light, then it would be perfectly clear what it would do: the light beam could be split into two by something called 'a half-silvered mirror' (a semi-transparent piece of glass that lets through half the light, which follows one path, and reflects the other half along a different path). This gives us two beams of light from the original single beam that we started with. These two beams, or light waves, travel along the separate routes through the device and combine again, interfering with each other at the end in a way that depends on the precise length of the path each has taken. If the two paths are exactly the same in length, the light waves will coincide – we say they combine 'in phase'; but if they arrive 'out of phase' we see destructive interference in some places, as though there were no light at all. The important thing to remember is that we only get this result where there are two waves combining.

Here is the real shocking feature of the quantum world. If a single electron is sent into an equivalent device, such that it is forced to choose between two paths (maybe by being made to deflect one way or the other by a magnet or an electrically charged wire) then, rather than doing what common sense tells us it should do, which is to go one way *or* the other, it will behave just like the light wave and somehow split up to *go down both paths at once*. How do we know it is doing this? Well, the result at the other end when the two possible routes are recombined is exactly what we would expect if the single electron were behaving like two spread-out waves travelling separately through the device.

Ever since the birth of quantum mechanics, physicists have been trying to figure out how particles such as electrons can do this. It seems they really are able to go along both paths at the same time; if they didn't, we would not see the wave-like behaviour of constructive and destructive interference that we do. It turns out that this is precisely what quantum theory predicts will happen: that we must describe quantum entities as waves when we are not looking. But as soon as we look, for instance by sticking some kind of detector along one of the paths in the interferometer, then we either observe the electron travelling along that path or we don't (implying it has taken the other path). So, when we check up on the electron while it is in transit we only ever see it going along one of the two routes. However, in doing so we will have unavoidably disturbed its quantum behaviour and any wave-like property of interference will disappear – which is not surprising, since now the electron is no longer travelling along both paths at once.

The lesson is this: in the quantum world, things behave in very different ways depending on whether we are observing them or not. When we aren't looking, they can be in a state of superposition, doing two or more things at once. As soon as we look, they are somehow immediately forced to make a choice between the various options and behave sensibly. The radioactive atom in the box with the cat really is in a superposition of two quantum states, having decayed and not decayed simultaneously. This is not because our ignorance means we have to 'allow' it the possibility of being in either state, but seemingly because it really is in a ghostly combination of the two.

The measurement problem

It is all very well having mathematical equations to describe the behaviour of atoms, but any half-decent scientific theory is only as good as the predictions it makes about the real world and the results of any experiment we care to set up to test those predictions. Quantum mechanics describes what goes on in the atomic world when we are not looking (which is a somewhat abstract mathematical description), and yet makes quite stunningly accurate predictions about what we would measure if we chose to. But the actual process of getting from the description of reality when we are not looking to the one we get when we point our measuring devices at it is still something of a mystery. It is known as the measurement problem. The issue is very straightforward to state: how do atoms and their ilk go from tiny localized particles to being spread out into multiple wavy versions of themselves, and back to behaving perfectly sensibly as tiny localized particles again as soon as we check up on them?

For all its success, quantum mechanics tells us nothing about how to take this step from the equations that describe how an electron, say, moves around an atom to what we see when we make a specific measurement of that electron. For this reason, the founding fathers of quantum mechanics came up with a set of ad hoc rules as an addendum to quantum theory. They are known as the 'Quantum Postulates' and provide a sort of instruction manual on how to translate the mathematical predictions of the equations into tangible properties we can observe, such as where an electron is located at any given moment.

As for the actual process itself whereby an electron instantaneously goes from being 'over here *and* over there' to just being 'over here *or* over there' when we look, no one really knows, and most physicists have been happy to adopt the pragmatic view laid down by Niels Bohr himself, who said that it just happens. He called it 'an irreversible act of amplification' and it is quite incredible to think that for most practising quantum physicists in the twentieth century this was enough. Bohr made an arbitrary distinction between the quantum world, where weird stuff is allowed to happen, and our much larger macroworld where everything behaves sensibly. The measuring device that looks at the electron has to be part of our macroworld. But *how* and *why* and *when* this measurement process takes place was not clarified. This was Schrödinger's problem: where is the dividing line between the micro and the macro? Presumably it must be somewhere between atoms and cats, but if so, how do we make the distinction if cats are themselves no more than a collection of atoms? In other words, any measuring device, whether it is a Geiger counter, an interferometer, an elaborate machine with lots of knobs and dials, or even a cat, is ultimately composed of atoms too. So where are we supposed to draw the line between the quantum domain, where the quantum rules are obeyed, and the macroworld of measuring devices? Come to think of it, what exactly constitutes a measuring device?

In our everyday world of large objects, we take for granted the notion that the way something appears to us is the way it really 'is'. But to see something means light has to reach our eyes from it. However, the very act of shining light on something we wish to see will disturb it and thereby alter it in some minute way when the light collides and bounces off it. This is not an issue when we look at something large like a car or chair or a person, or even a living cell under a microscope, where the collision of the particles of light (photons) with the object observed is not going to have any effect we could ever detect. But when we deal with quantum objects, which are themselves on the same scale as the photons, things are different. After all, every action has an equal and opposite reaction. So, in order to 'see' an electron we would have to bounce a photon off it. But in doing so, we will have knocked the electron off its original path.

In other words: to learn something about a system we must measure it, but in so doing we often unavoidably change it and will therefore not see its true nature. I have described this idea here in simple terms that do not really do justice to the subtlety of quantum measurement, but I hope it gives you some idea.

Let us pause for breath for a moment and review where we have got to. We know the quantum world is slippery and sneaky; not only can it get up to things that seem impossible in our everyday world, but it is devious enough to forbid us from ever catching it in the act of doing them. Opening up Schrödinger's box will always reveal a dead or a live cat, never a superposition of both. So it seems we are really no closer to resolving the paradox.

Desperate attempts

How, then, did physicists react to Schrödinger's paper? Bohr and Heisenberg themselves did not claim that the cat really was both dead and alive at the same time before the box was opened. But, rather than offer what might be regarded as a reasonable solution to the paradox, they wriggled out of it with a clever argument. They insisted that we cannot say anything about the cat or even give it an independent reality before we open the box to check up on it. Asking if the cat is really dead and alive at the same time is not an appropriate question.

Their reasoning was that all the time the box is closed we simply have nothing to say about the 'real' state of the cat. All we have to go by is what the equations predict we will find *when we open the box*. Thus quantum mechanics cannot tell us what is going on in the box, nor even what we will find when we open it. It can only predict the likelihood of finding the cat dead or alive. If such an experiment were to be carried out for real and repeated many times over (sacrificing many real cats), it would become apparent that such predictions are correct (just as we would need to flip a coin many times to confirm the fifty–fifty statistical probability of heads versus tails). These quantum probabilities are remarkably accurate, but we are only able to work them out if we insist that the atom is in this superposition of two states.

Over the years many physicists have tried, if not to explain away quantum weirdness, at least to come up with a way of getting to grips with just how the quantum world does what it does; and some of the most exotic suggestions have come up in response to the conundrum of Schrödinger's cat. One idea, known as the transactional theory, involves not only instantaneous connections across space, which is serious enough, but connections across time too. According to this view, the act of opening Schrödinger's box sends a signal into the past to tell the radioactive atom to 'decide' whether or not to decay.

At one time it was even fashionable to talk about a measurement requiring human consciousness that would force the quantum world into the macroworld – to suggest that there was something unique about consciousness which compelled the 'irreversible act of amplification' to take place and the quantum superpositions to disappear. After all, since no one knew where to place the boundary between the quantum domain of superpositions and the macro domain of definite outcomes when measurements are made, then maybe we should only place a boundary when we really have to. Since even the measuring apparatus (detector, screen, cat) is just a collection of atoms and should behave like any other quantum system, albeit a very big one, then we are forced to abandon the quantum description only when it registers in our conscious minds.

To place the dividing line between what is being measured and the measurer at the level of human consciousness amounts to what philosophers refer to as solipsism – the idea that the observer is at the centre of the Universe and that everything else is just a figment of his or her imagination. Thankfully, this view was largely discredited many years ago. What is fascinating, though, and not a little frustrating sometimes, is that there are still many non-physicists who argue that since we do not fully understand quantum mechanics or the origin of consciousness, the two must be related in some magical way. This kind of speculation, while fun, has no real place in serious science (yet).

What of the cat, then? Is it not conscious too? Can it not 'make an observation' while inside the box? There is an obvious way of testing this idea. What if we were to replace the cat with a human volunteer, and maybe the fatal poison with one that simply causes the volunteer to lose consciousness – which we could have insisted on for the cat too, I suppose? What happens now when we open the box? Clearly, we are not going to find the volunteer in a state of being simultaneously both conscious and unconscious, nor will we be able to convince him that he was in a superposition of both states before we let him out. If he is conscious he will report that, apart from feeling a little nervous, he has felt fine all along. And if we find him unconscious then he might, when revived, report how he heard the device go off ten minutes after the box was closed and immediately started to feel dizzy. The next thing he knew he was being revived with the smelling salts.

Thus, while an individual atom can exist in a quantum superposition, the volunteer clearly never did. And since there can be nothing special about our volunteer – does he have to have a PhD, or wear a white lab coat, for his consciousness to qualify for a measurement? – we would not be able to find any other sensible dividing line between him and the cat. We are thus forced to conclude that there is no reason why the cat should be described as being dead and alive at the same time before we open the box, if for no other reason than that the cat itself would know.

Quantum leakage

If the cat is never in a superposition of different states, then presumably the dividing line between the quantum microworld and our macroworld is further down towards the quantum end of the scale. Let's take another look at the question of what we mean by 'measurement'.

Consider what happens to an atom of uranium sitting inside a rock buried deep within the Earth. Very rarely, such atoms can spontaneously fission, splitting into two fragments that fly apart, releasing large amounts of energy. This is the energy utilized in nuclear reactors to provide heat, and hence electricity. These fragments of atomic nuclei, each roughly half the size of the original nucleus of the uranium atom, will emerge back to back, but fly apart in any direction. Quantum mechanics tells us that before a measurement is made, we must consider each fragment as having flown off in all directions. This is much easier to understand if we think of them not as particles but as waves, similar to the circular water waves spreading out from the pebble dropped into the pond. But we know that these fission fragments will leave tiny tracks in the rock that are visible in certain minerals under a microscope. In fact, studying these tracks, just thousandths of a millimetre in length, is a useful technique in the radiometric dating of rocks.

So, the issue is this: since these tracks are produced in the quantum world, until we measure them we must describe them as having both appeared, if the uranium nucleus has fissioned, and not appeared, if it has not. And if it has, we should describe them as having emerged in all directions at once. But what then constitutes a measurement? Does the rock sit in limbo, with tracks both etched and not etched in it, until we look at it under the microscope? Of course not; they will either be there or not, whether we analyse the rock today, in a hundred years' time or never.

Measurements of the quantum world must be taking place all the time, and conscious observers, whether they wear lab coats or not, cannot play a role. The correct definition is that a measurement is deemed to have taken place when an 'event' or 'phenomenon' is recorded, in the sense that a trace of the event has been left that we could perceive, if we wished to do so, at a later time.

This might sound so obvious to you that you could be forgiven for wondering how quantum physicists could be so stupid as to think anything else. But then again, some of the predictions of quantum mechanics are anything but rational. What is needed is a clearer idea of how events in the quantum world get recorded, for that is when the quantum weirdness (going in two directions at once, or both doing and not doing something simultaneously) must leak away.

During the 1980s and 1990s, physicists came to appreciate what must be going on. They considered what happens when an isolated quantum system, such as a single atom, ceases to exist in its happy solitary superposition and becomes coupled with a macroscopic measuring device, which could even be its surrounding environment, such as the rock. Quantum mechanics dictates that the many trillions of atoms that make up the measuring device/rock must also exist in a superposition. However, these delicate quantum effects become too complex to be maintained in such a large macroscopic device and leak away, almost like heat dissipating from a hot body. The process is known as 'decoherence' and is currently a subject of much debate and research. One way to think about it is to say that the individual delicate superposition gets irretrievably lost among the stupendously large number of other possible superpositions arising from the different possible combinations of interactions between all the atoms in the macroscopic system. Recovering the original superposition is a little like trying to un-shuffle a pack of cards, only far more difficult.

Many physicists today regard decoherence as a real physical process that is going on everywhere in the Universe all the time. It takes place whenever a quantum system is no longer isolated from its surrounding environment (which can be anything from a Geiger counter to a lump of rock or even surrounding air molecules; it certainly does not have to involve a conscious observer). If the connection to its external environment is strong enough, then the original delicate superposition gets lost very quickly. In fact, decoherence is one of the fastest and most efficient processes in the whole of physics. And it is this remarkable efficiency that is responsible for decoherence having evaded discovery for so long. It is only now that physicists are learning how to control and study it.

Even though decoherence is not yet fully understood, we can at least begin to make sense of our paradox. The reason we never see Schrödinger's cat both dead and alive at the same time is because decoherence takes place within the Geiger counter long before we open the box. Because of its ability to register whether or not the atom has decayed, the Geiger counter forces the atom to decide. So, during any given interval of time, the atom will either have decayed or not decayed and the Geiger counter will either have registered it, setting in motion the sequence of events that kills the cat, or it won't. Once we have emerged from the quantum world of superposition, there is no going back and we are left with simple statistical probability.

A neat experiment carried out and reported in a paper published in 2006 by two Cambridge scientists, Roger Carpenter and Andrew Anderson, confirmed that the collapse of the superposition and leaking away of quantum weirdness does indeed take place at the level of the Geiger counter itself. The experiment received very little attention, maybe because most quantum physicists believe there is no puzzle to resolve any longer.

It seems, therefore, that decoherence tells us not only why we never see Schrödinger's cat both alive and dead at the same time, but why the cat itself never exists in this in-between state in the first place. What decoherence doesn't tell us, of course, is how one or the other option is selected. Quantum mechanics remains probabilistic, and this unpredictability of individual measurements does not go away.

In fact, even this need to explain how one or other of the two possible options gets selected does not need explaining if one subscribes to the Multiverse theory. Here, the cat will be dead in one universe and alive in another. When you open the box, all you are doing is finding out which universe you are in, the dead cat one or the live cat one. Whichever universe you find yourself in, there will always be another version of you in another universe opening the box to find the alternative outcome. Simple, really.

TEN

Fermi's Paradox

Where is everybody?

T HE ITALIAN AMERICAN Nobel Prize winning physicist Enrico Fermi made many important contributions to quantum mechanics and atomic physics. He built the very first nuclear reactor, Chicago Pile-1, in the early 1940s; he has one of the two classes of elementary particles, the 'fermion', named after him (the other being the boson); there is even a unit of length that bears his name – the 'fermi', which is the alternative name for the very tiny length of one femtometre, or a trillionth of a millimetre, the appropriate scale for use in nuclear and particle physics. But this chapter is about a question Fermi posed in 1950 that had no connection with his research in subatomic physics. It concerns the most profound and important paradox of them all – and I've saved it till the very end.

Fermi's famous question came up during a lunchtime conversation he had with a number of colleagues during a summer visit to the Los Alamos National Laboratory in New Mexico, home of the atomic bomb and the Manhattan Project. The conversation had revolved around some light-hearted debate over flying saucers and whether they could possibly exceed the speed of light in order to reach the Earth from distant star systems.

Fermi's Paradox can be stated in the following way:

Since the age of the Universe is so great and its size so vast, with hundreds of billions of stars in the Milky Way alone, many of which have their own planetary systems, then unless the Earth is remarkably atypical in having the conditions to harbour life, the Universe should be teeming with it, including intelligent civilizations, many of which would have the technology required for space travel and would have visited us by now.

So, where is everybody?

To Fermi it was obvious that, assuming our solar system is not unique in containing (at least) one habitable planet, there has been plenty of time for any alien civilization with even modestly expansionist ambitions and a sufficiently well-developed space travel technology to have colonized the entire galaxy by now. He and others have estimated that it would take 10 million years for any species to do this. While this may seem like a very long time, and is certainly a somewhat arbitrary figure, the important point to note is that it is a tiny fraction, in this case just one-thousandth, of the age of the galaxy – and remember that *Homo sapiens* has only been around for about 200,000 years.

The paradox can therefore be boiled down to the following two questions:

- If life is not so special, where is everybody else?

- If it is, then how come the Universe is so finely tuned as to allow life to emerge only on Earth?

If we think about the ability of life on our own planet to proliferate and flourish in the harshest of environments, then why shouldn't it have done the same thing on other Earth-like planets? Maybe the problem is not with the proliferation of life once it gets started, but its getting started in the first place. Before we examine whether scientists have been able to resolve this paradox and the various issues involved, let us briefly look at some of the solutions that are commonly suggested.

1 **Extraterrestrials exist and have in fact already visited us**. I will dismiss this first option for the justifiable reason that we have no sensible evidence to support the fantastical delusions of UFO enthusiasts and conspiracy theorists. Despite this, many people remain convinced that aliens have already arrived in flying saucers, whether pausing thousands of years ago to build the pyramids before leaving again, or remaining here today, abducting innocent victims in order to conduct bizarre experiments on them.

2 **Extraterrestrials are out there but have not got in touch**. There are plenty of reasons we can think of why sufficiently advanced alien civilizations have chosen not to give us any indication of their existence. For example, perhaps (unlike us) they do not wish to broadcast their existence to the rest of the galaxy, or perhaps they prefer to leave us alone until we become advanced enough to qualify for membership of the galactic club. This assumes, of course, that all alien civilizations follow similar reasoning processes to ours.

3 **We're not looking in the right place**. Although we have been listening out for signals from space for fifty years, we have yet to hear anything at all. But maybe we have yet to look in the right region of the sky, or tune in to the right frequency; or maybe signals and messages have already arrived but we have not figured out how to decode them.

4   **Life elsewhere is regularly destroyed**. We may not appreciate just how privileged we are on Earth. Life-supporting planets in other solar systems may have to endure a whole range of catastrophic planetary, stellar or galactic events on a regular basis, such as ice ages, meteor or comet impacts, massive stellar flares or gamma-ray bursts. Where such events happen frequently, life wouldn't have time to evolve intelligent, spacefaring species. Or maybe the opposite is true, and the environment on other planets is so comfortable that they have not endured the mass extinctions thought necessary to encourage biodiversity and thus the evolution of intelligence.

5   **Self-destruction**. It has been suggested that all intelligent life in the Universe will inevitably annihilate itself, either through war or disease or by destroying its environment, around about the time it becomes technologically advanced enough for space travel – which, if true, is an ominous message for us.

6   **Aliens are just too . . . alien**. We tend to assume that extraterrestrials would be like us, with technologies similar to those we can envisage developing in the future. While there are good reasons for thinking like this, since all life would have to abide by and be constrained by the laws of physics, it may be that we just don't have the imagination to conceive of intelligent life sufficiently different from us. I don't of course mean that we think they will all look like the aliens in the movies, but we do tend to assume they will be carbon-based, have limbs and eyes, and communicate by sending each other sound waves.

7   **We truly are alone in the Universe**. Maybe the conditions necessary for any life to emerge are so rare that it has only happened in very few places, and the Earth is the only planet where intelligent life has evolved capable of harnessing nature

in order to send out signals into the Universe announcing its existence. Or maybe our planet really is the only place where life has appeared at all.

All of the above scenarios are just guesswork, and mostly not very educated guesswork at that. Fermi's own view was that although it was overwhelmingly likely that intelligent life did indeed exist elsewhere in the galaxy, the distances involved in interstellar travel were so vast, and would take so long to traverse, given the speed of light barrier, that no civilization would deem it worth the effort to visit us.

What Fermi did not take into account was the fact that we might be able to discern the existence of technologically advanced extraterrestrials even if they never left their home planet. After all, we have been announcing our presence to the galaxy for almost a century. Ever since we began to use radio and television to transmit information around the world we have been leaking signals out into space. An alien civilization a few tens of light-years away that happened to point its radio telescopes at our Sun would pick up an extraordinary amount of faint yet complex radio signals that would be a sign of life on one of the planets orbiting it.

Given that we believe the laws of physics to be the same throughout the Universe, and given that one of the easiest and most versatile means of transmitting information is by using electromagnetic waves, we would expect any technologically advanced civilization to make use of this form of communication at some point in its development. And if it does, some of its signals will leak out into space, spreading through the galaxy at the speed of light.

It wasn't long before twentieth-century astronomers began seriously to consider the feasibility of listening for signals from space using their newly developed radio telescopes. And the serious search for extraterrestrial intelligence began with one man.

Drake and his equation

The first serious ET hunter was the astronomer Frank Drake, who worked at the National Radio Astronomy Observatory in Green Bank, West Virginia. In 1960 he set up an experiment to search for signs of life in distant solar systems by listening in to electromagnetic signals at radio-wave frequencies. The project was named Ozma, after Princess Ozma, ruler of the Emerald City of Oz in Frank Baum's children's books.

Drake pointed his radio telescope at two nearby Sun-like stars, Tau Ceti and Epsilon Eridani, respectively twelve and ten light-years away, both of which seemed reasonable candidates to host habitable planets. He tuned his dish to pick up radio signals at a particular frequency: that of the quite specific electromagnetic radiation produced by the lightest, simplest and most abundant element in the Universe, and therefore the most likely choice for any alien civilization trying to make itself known: hydrogen. He recorded the data and carefully checked to see whether there was any signal superimposed on to the general hiss of background noise. After many hours of recorded data over several months had been examined, no interesting signals were found apart from one, which turned out to have come from a high-flying aircraft. But Drake was not disappointed. He has always maintained that the process was like buying a lottery ticket: he knew he would have been incredibly lucky if something had been found.

Undeterred, the following year he organized the very first conference on SETI (Search for Extraterrestrial Intelligence) and invited along every other scientist he knew to be interested in the subject at the time (all twelve of them).

In order to focus their minds, he devised a mathematical formula for calculating the number of civilizations, N, in our galaxy whose radio signals would be detectable on Earth. He calculated this by multiplying seven other numbers together. The formula, which now bears his name, looked like this –

$$N = R_* \times f_p \times n_e \times f_l \times f_i \times f_c \times L$$

– and it is actually quite straightforward to explain. I will run through what each of the symbols means and, in each case, I have included in brackets the value Drake assumed in his first calculation so you can see how he arrived at his final number. The first symbol, R_*, stands for the average number of new stars forming in the galaxy every year (Drake assumed this to be 10 per year). The next, f_p, is the fraction of those stars with planetary systems (0.5); n_e is the number of planets, per solar system, with an environment suitable for life (2); f_l, f_i and f_c represent, respectively, the fraction of suitable planets on which life actually appears (1); the fraction of life-bearing planets on which intelligent life emerges (0.5); and the fraction of those civilizations that develop a technology that releases into space detectable signs of their existence (1). Finally, L is the length of time over which such civilizations would continue to release detectable signals into space (10,000 years). Multiplying these seven numbers together, Drake arrived at an answer of $N = 50,000$.

This is an impressive number and serves to highlight Fermi's Paradox. But how reliable is it? The answer, of course, is: not very reliable at all. Even if these seven quantities really were all we needed to know, the values accorded to a number of them are no more than wild guesses. The first three, R_*, f_p and n_e, are set by values that, while not known half a century ago, are beginning to become clearer with advances in astronomy and telescope technology, particularly since many planets outside our solar system, known as extrasolar planets, have recently been discovered.

However, the next three factors are probabilities to do with the likelihood of intelligent, communicative life forms emerging. Each of them could have a value of pretty much anything between 0 (impossible) and 1 (certain). Drake chose extremely optimistic values; he believed that if the conditions were right on some Earth-like planet then the appearance of life was inevitable (f_l=1), that if life did emerge then there was a fifty–fifty chance that it would evolve intelligence (f_i=0.5), and that if it did, then this intelligent species would surely develop technology that involved electromagnetic waves that would be emitted into space, whether it was deliberately sending a message or not (f_c= 1).

But the numerical values are almost by the way. What Drake's equation did was something far more important than provide an estimate of the number of alien civilizations out in the galaxy. It kicked off a worldwide hunt for signals from space that continues to this day.

SETI

SETI is the collective name for a number of projects around the world that have been conducted over the years to search actively for extraterrestrial signals. Listening out for potential messages from space transmitted via electromagnetic waves is something we have done ever since scientists began to understand how to send and receive such signals. One of the earliest incidents goes as far back as the end of the nineteenth century.

In 1899, while investigating atmospheric electricity from storms in his Colorado Springs laboratory using his newly developed and highly sensitive radio frequency receiver, Serbian-born electrical engineer and inventor Nikola Tesla detected faint signals coming in clusters of what appeared to be a numerical code of one, two, three and four 'beeps', which he was convinced had originated on Mars. He recalls his excitement in a magazine interview in 1901:

I can never forget the first sensations I experienced when it dawned upon me that I had observed something possibly of incalculable consequences to mankind . . . My first observations positively terrified me, as there was present in them something mysterious, not to say supernatural, and I was alone in my laboratory at night . . . [The electrical signals arrived] periodically and with such a clear suggestion of number and order that they were not traceable to any cause known to me . . . It was sometime afterward when the thought flashed upon my mind that the disturbances I had observed might be due to an intelligent control.*

Although Tesla was widely criticized for his claims, the mystery of the signals he detected remains unsolved.

The first serious investigation into possible radio signals from intelligent extraterrestrials was a short-lived project that took place in America in 1924. At the time it was still believed that the most likely home planet for an alien civilization was our near neighbour, Mars; and that, if Martians were going to communicate with us, then they would do so when the two planets were closest together. This happens during what is called an 'opposition', when the Earth passes between Mars and the Sun. One such opposition took place between 21 and 23 August 1924, when Mars was closer to the Earth than it had been for thousands of years (a record that was beaten in August 2003 and will be again in the year 2287). It was decided that if Martians existed then they would use such an opposition event to transmit signals to Earth. The United States Navy took this idea seriously enough to hold a 'National Radio Silence Day', asking for all radios around the country to be turned off for five minutes on the hour, every hour, during the

*'Talking with the Planets', *Collier's Weekly*, 19 Feb. 1901, pp. 4–5.

36-hour period as Mars passed by. At the US Naval Observatory in Washington DC, a radio receiver was taken up to 10,000 feet in an airship and all naval stations around the country were instructed to monitor the airwaves for anything unusual. All they heard was static – along with the signals from those private broadcasters who didn't observe the radio silence.

It was following Frank Drake's initial projects that the SETI movement really took off, extending its search far beyond the solar system. To give you an idea how much further radio telescopes had already extended our listening range, the two stars that Drake focused on in 1960 are about ten light-years away – 2 million times as far away as Mars. It's a bit like sticking a cup against your wall to listen to your neighbour's conversation, not hearing anything, and deciding instead to try to listen to a conversation in New York when you are in London. Clearly, choosing precisely where to point the radio telescope dishes was crucial.

The SETI Institute in California was set up in 1984 and several years later began Project 'Phoenix' under the direction of astronomer Jill Tarter, who is the inspiration for the main character of Carl Sagan's novel *Contact*. Between 1995 and 2004 Project Phoenix used radio telescopes in Australia, the US and Puerto Rico to look at 800 Sun-like stars within 200 light-years of Earth. They found nothing. But the project has produced a valuable source of information for research into possible alien life. Together with fellow astronomer Margaret Turnbull, Jill Tarter put together a catalogue of nearby stars that might have planetary systems capable of harbouring life, called 'habstars' (habitable stars). The catalogue, known as HabCat, currently contains over 17,000 stars, most of which are within a few hundred light-years of the Earth and possess the right characteristics and qualities to make them candidates for hosting Earth-like planets.

In 2001 Microsoft co-founder Paul Allen agreed to fund the initial phase in the construction of a new radio telescope array dedicated to

SETI. Called the Allen Telescope Array (or ATA), it is still under construction a few hundred miles north-east of San Francisco. When complete, it will consist of 350 radio dishes, each 6 metres in diameter, working in combination. The first phase was completed in 2007, when forty-three of the antennas became operational, but the project was temporarily halted in early 2011 owing to government research funding cuts. Soon after, a group was set up to try to save it by seeking private funding from anyone who wished to help. Thousands signed up to provide donations, including the film star Jodie Foster, who played the role of Jill Tarter's character in the Hollywood version of Carl Sagan's *Contact*. For some reason I find all this rather sweet and satisfying.

Far from giving up, then, the search for ET is only now beginning in earnest. To date, we have looked carefully at only a few thousand stars in a limited range of frequencies of the full electromagnetic spectrum. The plan for the ATA is to survey 1 million stars out to 1,000 light-years. The frequency range of the search is also being widened. Drake's initial choice of the frequency of interstellar hydrogen, 1.42 GHz, was a sensible one. Our skies are very noisy places – there are radio waves coming from all sorts of sources, including galactic noise and the noise of charged particles moving through the Earth's magnetic field, as well as the cosmic background radiation left over from the early universe. But the range of frequencies to be scanned by the ATA, lying between 1 and 10 GHz and known as the 'microwave window', is a particularly quiet region of the electromagnetic spectrum and so ideally suited for searching for extraterrestrial signals.

More serious academic research, however, has focused in recent years on searching not for signs of intelligent life but for Earth-like planets that might host it. Today, extrasolar planet-hunting is one of the hottest areas of scientific research.

Exoplanets

I am sure that I'm not alone in finding the search for and study of extrasolar planets (or exoplanets) incredibly exciting. Observing and studying stars is one thing – after all, we can tell a lot about their composition and how they're moving from the light they give off. But planets are an entirely different matter. Not only are they much smaller than stars, they only give off their stars' reflected light and therefore shine a million times less brightly than even the dimmest star. So any evidence of their existence can usually be inferred only indirectly. The most common technique is the so-called transit method, whereby a tiny dip in the brightness of a star is detected when a planet passes in front of it. Another method is to observe the small effect a planet's gravity can have on its much more massive star, causing it to wobble slightly. This can be picked up either as a change in the frequency of the star's light as it moves towards us or away from us (the Doppler shift) or by directly measuring changes in its position.

Of particular interest to astronomers are those planets that are Earth-like in the sense of being solid and having roughly Earth-like gravity, as well as being the right distance from their star to allow for liquid water to exist on their surface, thus allowing them, potentially, to host life.

At the time of writing, about 700 extrasolar planets have been found. But this figure is likely to rise fast. In 2009 NASA's Kepler mission launched a spacecraft equipped with the instruments necessary to discover exoplanets. In February 2011 Kepler's research group announced a list of 1,235 possible exoplanets, including fifty-four that looked to be in the 'habitable zone'. Of these, six are Earth-sized or nearly so.

It has been estimated that there are at least 50 billion planets in the Milky Way, at least 1 per cent (500 million) of which are in the habitable zone. Other estimates put this figure of habitable Earth-like planets at over 2 billion. Of these, as many as 30,000 are within 1,000 light-years of Earth.

So far, two confirmed habitable zone exoplanets in particular have captured the scientific community's imagination, not because they have shown any evidence of supporting life, but because they are the closest candidates yet to what have been referred to as 'Goldilocks planets'; that is, having all the right conditions for life, being neither too hot nor too cold – just like Baby Bear's porridge in the children's fairy tale. The first is called Gliese 581d and orbits its parent red dwarf star, Gliese 581, 20 light-years from Earth in the constellation of Libra. Note the letter 'd' at the end of its name, indicating that it is the third planet to have been discovered orbiting this star (the planets belonging to any star are named by the sequence of letters running in alphabetical order starting from 'b' – the star itself being 'A'). While Gliese 581d is over five times the size of the Earth, recent climate simulation studies suggest it has a stable atmosphere and liquid water on its surface. Several other potentially habitable planets have also been discovered orbiting the same star, but they have yet to be confirmed.

The second candidate is HD85512b, orbiting the star HD85512 (so named because it is catalogued in the Henry Draper star catalogue), which is 36 light-years away from us in the constellation of Vela. This is one of the smallest habitable zone exoplanets so far discovered and is currently regarded as the best candidate for hosting alien life. It is about four times the size of the Earth with a surface gravity of about one-and-a-half times that of our own and an estimated temperature of 25°C at the top of its atmosphere. The temperature on its surface is as yet unknown but is likely to be considerably higher. Its year – the time it takes to orbit its star – is just fifty-four days.

There was much excitement in late 2011 when the Kepler mission announced its first confirmed exoplanet, Kepler 22b. Although its parent star is significantly further from the Earth than both Gliese 581 and HD85512, at almost 600 light-years away, it is very similar to our own Sun (called a G-type main sequence star). Not much is yet known about exactly how big the planet Kepler 22b is, although current estimates put its diameter at several times that of the Earth; nor is it yet known whether it is a rocky planet like the Earth or a gas planet like Jupiter and Saturn. If it is indeed rocky then it is feasible that it could have liquid water on its surface; and the fact that it orbits at the right distance around a star like our Sun makes it a potentially exciting candidate for harbouring life.

Whether we are likely to find answers to all these questions any time soon is debatable; but we have come a long way in exoplanet research in a very short time and discoveries continue to come in thick and fast.

How special are we?

Of course, a planet's being suitable to support life is one thing, but the really big unknown is this: given the right conditions, how likely is it that life could evolve elsewhere? To answer that we need to understand how life began on Earth.

Our planet is teeming with life, from flora and fauna to bacteria. And many species, particularly microbes, seem to be able to thrive in the harshest of environments, from extreme cold to extreme heat, with and without sunlight. This diversity of life, and the fact that it seems to have taken hold relatively quickly after the young Earth cooled, suggests that it wasn't very difficult for it to get started. But is that a correct view? We now know that the right conditions exist to support at least bacterial life elsewhere in the Universe (or, more specifically, elsewhere in our solar system), so it seems only reasonable to expect that life might have appeared on other worlds too. But just how special is our home planet?

The Earth is certainly just the right distance from the Sun – not too hot, not too cold. It also benefits from having the giant planet Jupiter in orbit outside it, because Jupiter acts like a protective big brother, sucking up space debris with its massive gravitational pull and so preventing its reaching Earth's orbit and bombarding us.

The Earth's atmosphere is crucial, not only for providing us with the air we breathe – after all, life started on Earth before there was any oxygen in the atmosphere – but because of the way it interacts with electromagnetic radiation. It is transparent to visible light, but partially absorbs infrared light (heat) both on its way in (from the Sun) and on its way out (radiated from the Earth's surface). This 'greenhouse effect' serves to heat up the atmosphere and, in turn, keep the water on the planet's surface in liquid form, which is far more useful in encouraging life to flourish than either ice or vapour.

Our moon is vital too. Its gravity stabilizes the Earth's spin, giving it a settled climate for life to evolve, while the tidal forces it generates on the Earth's mantle as it orbits, particularly when it was very much closer to the Earth billions of years ago, may have helped heat the mantle and possibly also generate the Earth's magnetic field. This in turn shields our planet from the solar wind, which would otherwise blow away the Earth's atmosphere into space.

Even such processes as plate tectonics turn out to be crucial since they help to recycle the carbon needed to stabilize the temperature of the atmosphere and replenish the nutrition available to living creatures on the Earth's surface; they may also contribute towards the planet's magnetic field.

So perhaps our planet is rather remarkable after all. But does that mean life was inevitable here? Once life got started and evolution took over, life took care of itself, but the real issue is that very first step.

The first living things on Earth are thought to have been single-cell prokaryotes (simple organisms that lack a cell nucleus), dating back to about three and a half billion years ago. These may have in turn evolved from what are known as protobionts, which are no more than a collection of organic molecules enclosed in a membrane, but with the ability to replicate and metabolize, two of the key signatures of life.

What we don't yet know is what sequence of events could have led from organic molecules like amino acids (needed to form proteins) and nucleotides (the structural units of our DNA) to the very first 'replicator'. This question of how life began is one of the most important in science. It is known as abiogenesis. Many people make the mistake of confusing biogenesis (the theory that life can only emerge from other life) with abiogenesis (the process by which biological life emerges from inorganic matter through natural processes – essentially, how chemistry turns into biology). Research into abiogenesis seeks to find out what that magical step was – that step commonly known as 'spontaneous generation' – that turned inanimate matter into life.

It has been argued that the spontaneous emergence of life on Earth would have been such a rare event that it is the equivalent of a strong wind blowing through a junkyard and, from the materials there, creating an entire jumbo jet, by accident. That, the argument goes on, is how likely it would be for organic molecules to fit together through chance alone in just the right combination to make even the most basic life forms – quite an amazing level of coincidence. Is that a fair comparison?

In 1953 Stanley Miller and Harold Urey at the University of Chicago conducted a famous experiment to try to address this question. They wanted to see if they could create life in a test-tube from its basic ingredients. They mixed water with three gases – ammonia, methane and hydrogen – in the belief that this matched the atmosphere of the early Earth, and the concoction was heated up and evaporated. They then delivered sparks from two electrodes to the vapour to simulate lightning in the Earth's atmosphere, before condensing the vapour. After one week of continuously repeating this process, they noticed that organic compounds had begun to form, including amino acids, compounds vital for life since they are strung together in particular arrangements to make up proteins in living cells. But complete proteins in all their complexity were not seen; nor were the other vital life ingredients, nucleic acids (such as DNA and RNA).

Despite such a promising start, in over half a century since that key experiment scientists have yet to create artificial life. So is it really so improbable for life to start spontaneously? We know it happened at least once – we're the living proof of that – but it would be interesting to know whether all life on Earth today has one single ancestor, because if it didn't then this would mean life had to start on more than one occasion and is therefore less special than we may think.

A recent controversial piece of research seemed to challenge this idea. It came with the discovery of a new microbe by the name of 'strain GFAJ-1' (proving that microbiologists are just as unimaginative as astronomers in naming their discoveries) in a strange desert lake in California. Mono Lake, which formed about a million years ago, has very unusual chemistry. It is two to three times as salty as the ocean, containing chlorides, carbonates and sulphates, and is very alkaline, with a pH of 10. Although it contains no fish, the chemical mix of the lake's water makes it an ideal habitat for single-celled algae of a certain type and trillions of tiny brine shrimp that provide staple food for millions of migratory birds that gather there for several months each year. Oh, and the lake is also rich in arsenic.

A team of NASA biologists led by Felisa Wolfe-Simon became interested in the tiny GFAJ-1 bacterium, because it seemed to be able to do something never seen before: it fed on the arsenic, an element poisonous to all other life.

We know that life on Earth depends on a mix of many different elements, but DNA itself is made up of just five ingredients: carbon, hydrogen, nitrogen, oxygen and phosphorus. The question is whether other elements chemically similar to these can be substituted for them. Arsenic sits just underneath phosphorus in the periodic table and so has a similar atomic structure. The NASA researchers knew that GFAJ-1 was arsenic-tolerant and that Mono Lake contained very little phosphorus. So they put it on an arsenic-rich diet, and it thrived – even when they removed the phosphorus entirely. Now, when cells replicate they need the raw materials to build new DNA, so how were these organisms coping without one of the five crucial ingredients?

The researchers' publication of their work at the end of 2010 created a worldwide storm in the scientific community. They claimed that GFAJ-1 was literally substituting arsenic for phosphorus in the very structure of its DNA. If this turns out to be true, then we are faced with the million-dollar question: Did these microbes evolve the ability to utilize arsenic in this way, or did they emerge in a separate abio-genetic event? If the latter, then we would know that life could have started on at least two separate occasions – and thus that it may not be so rare.

We still do not know how life got started on Earth. Even if and when we can answer this question, there are yet other puzzles surrounding the likelihood of the emergence of *intelligent* life. After all, it may be that life exists in many locations around our galaxy, but intelligent life exists in just one place.

Recent research studying the behaviour of crows has suggested that these birds have evolved remarkable intelligence along a completely separate evolutionary path from the one that led to humans. If so, then maybe intelligence is also an inevitable result of

Darwinian evolution. This and other issues, such as the way multi-celled organisms evolved from single-celled ones billions of years ago, will tell us whether we can expect these important steps in the evolutionary journey from abiogenesis to humans to have occurred elsewhere in the Universe.

The anthropic principle

There exists a far more profound question than the one posed by Fermi's paradox that I must mention before I finish this chapter. It is one that until recent years was restricted to philosophical circles alone, but has now entered mainstream physics. The idea at its core is called the anthropic principle, which focuses on the sheer improbability of our universe, or at least our small corner of it, being so ideally suited and fine-tuned for us, humans, to exist. In its modern form it was proposed and clarified by the Australian cosmologist Brandon Carter at a scientific conference held in Poland in 1973 to celebrate the 500th anniversary of the birth of Copernicus. Carter stated it in the following way: 'What we can expect to observe must be restricted by the conditions necessary for our presence as observers. Although our situation is not necessarily *central*, it is inevitably privileged to some extent.' This was a particularly fascinating occasion to introduce such an idea since Copernicus had been the scientist who first proposed that humankind did *not* occupy a privileged position in the Universe. Now here was Carter suggesting that the whole Universe looks the way it does to us because had it been any different we wouldn't have even existed. Let me give you an example from my own field of nuclear physics. One of the four fundamental forces of nature is the strong nuclear force, which is responsible for binding atomic nuclei together. Two hydrogen nuclei (single protons) will not bind because the strong nuclear force isn't quite strong enough to do that – but it is strong enough to stick a proton and a neutron together to produce something called a deuteron (the nucleus of a 'heavy hydrogen' atom), which plays a vital role in the sequence of steps in nuclear fusion that turn hydrogen into helium, the process that drives all stars and provides us with the life-giving light and warmth of our Sun. But what if the strong nuclear force were just a fraction stronger? It might then be strong enough to bind two protons, in which case hydrogen would convert to helium far more easily. In fact, all the hydrogen in the Universe would have been used up and converted into helium just after the Big Bang. With no hydrogen, there would be no possibility of combining it with oxygen to make water and hence no chance of life (as we understand it) ever appearing.

The anthropic principle seems to be saying that our very existence determines certain properties of the Universe, because if they were any different we would not be here to question them. But is this really so remarkable? Maybe if the Universe were different, we – whatever 'we' would then mean – would have evolved according to whatever those conditions would have allowed and we would still be asking: how come the Universe is so finely tuned?

One way of thinking about this is to ask yourself: how come you, personally, exist? After all, what were the chances that your parents would meet and produce you? And what were the chances of their parents producing them, and so on? We are each at the end of a long and highly improbable chain of events that stretches all the way back to the origin of life itself. Break any one of the links in that chain and you would not be here. So you can ponder if you wish how the anthropic principle applies to you; but this is no more remarkable than the lottery winner contemplating his good fortune. And had his numbers not come up, then someone else would have won and could equally reflect on the improbable odds of her win.

Brandon Carter's argument has become known as the Weak Anthropic Principle. There also exists a Strong Anthropic Principle, which states that the Universe has to be the way it is in order for intelligent life to evolve somewhere and at some point in time in order to question its existence. This version is subtly different and is far more speculative. Personally, I think it is nonsense. It imputes a purpose to the Universe in arguing that it has somehow been compelled to turn out the way it has in order to produce us. There is even a quite sophisticated quantum mechanical version of this argument that has parallels with the 'conscious observer' solution to the Paradox of Schrödinger's Cat: by our observing the Universe we have brought it into existence, all the way back in time. Out of all the potential universes, we 'selected' the one that allows us to exist in it.

A much simpler way out of the anthropic puzzle can be found if we just surrender to the seductive charm of the multiverse theory. After all, if every possible universe exists then it is no surprise that we find ourselves in one that is just right for us.

*

Let me end this chapter by returning to where I began it, with the famous question asked by Enrico Fermi about the eerie silence from space. After all, a finely tuned universe for us would also be a finely tuned universe for other life forms not too dissimilar from us. While the vastness of the Universe with its billions of galaxies suggests that, however special the Earth is and however unlikely the emergence of life on it was, it is overwhelmingly likely that life exists elsewhere, it may simply be that we are alone in our little corner of the Milky Way.

Why then do we continue to search, possibly in vain? It is because we seek answers to the most fundamental questions of existence. What is life? Are we unique? What does it mean to be human and what is our place in the Universe? Even if we never find answers to these questions, it is important that we continue to ask them.

ELEVEN

Remaining Questions

*Can particles go faster than light? Do we have free
will? And other remaining conundrums*

I HOPE YOU WILL agree with me that we have successfully confronted
and banished nine of the very best scientific paradoxes. We have
exorcised demons, rescued cats and grandfathers, stopped squabbling
twins, made our peace with the night sky and put Zeno the Greek in
his place. But you might be wondering whether I have been selective
in tackling only those puzzles that have been neatly solved by science,
and whether there are still others out there which I have conveniently
ignored because we don't yet have answers to them. Well, of course
there are. The Universe is still full of mystery – that's what makes it so
fascinating.

All such remaining conundrums and mysteries can be put into
one (or more) of three categories: those that science is on the verge of
understanding and solving; those that science hopes to solve one day,
possibly far in the future; and those philosophical or metaphysical
problems that science may never answer, either because they fall
beyond its remit or because we cannot, even in principle, find any
conceivable way of investigating them in order to tease out a definitive
and satisfying answer.

Rather than describe these outstanding problems in science in any detail at this late stage of the book, I will merely group some examples into the different categories. I should stress that I don't rank them in any particular order of how close I believe they are to being resolved. I should also stress that these are personal and highly subjective lists, neither comprehensive nor restricted to questions and puzzles that are paradoxical. I set them out here merely to highlight how much more we still need to learn about the Universe and our place in it.

I begin with ten problems that fall into the first category – those to which I anticipate science will find satisfactory answers within my lifetime:

1   Why is there more matter than antimatter in the Universe?

2   What is dark matter made of?

3   What is dark energy?

4   Will fully functional invisibility cloaks be possible?

5   How far can we push chemical self-assembly towards explaining life?

6   How does a long strand of organic material fold up into a protein?

7   Is there an absolute limit to human life spans?

8   How are memories stored and retrieved in the brain?

9   Will we one day be able to predict earthquakes?

10   What are the limits of conventional computing?

Next come ten outstanding problems I am confident science will answer one day, though I doubt I will live to see it happen:

1   Are particles really tiny vibrating strings or is string theory just clever maths?

2   Was there anything before the Big Bang?

3   Are there hidden dimensions?

4   Where and how does consciousness originate in the brain?

5   Can a machine ever be conscious?

6   Is time travel to the past possible?

7   What shape is the Universe?

8   What is at the other end of a black hole?

9   Do deeper principles underlie quantum weirdness?

10   Will quantum teleportation of humans ever be possible?

Finally, here are a few problems that many would argue are in principle within the remit of science, but which I fear science may never be able to answer:

1   Do we have free will?

2   Are there parallel universes?

3   What caused the Universe to come into existence?

4   Did we invent mathematics to describe the Universe or were the equations of physics always out there just waiting to be discovered?

Faster than light?

Before I wrap up this final chapter, I would like to give you an example of what many would regard as a potential paradox if a recent experimental result is to be believed. At the time of writing there are two as yet unresolved puzzles in particle physics that made headlines around the world in 2011; both are being addressed by experiments carried out at the particle accelerator in CERN, Geneva. The first is whether particles can travel faster than the speed of light; the second is whether the elusive Higgs boson, the elementary particle that gives substance to the Universe, actually exists. In both cases, results to date have been inconclusive, and in both cases further experimental work is required. In an effort to endow this book with reasonable longevity I will stick my neck out and offer a prediction about how these two issues will be resolved. The Higgs boson will be confirmed to exist in the summer of 2012; and subatomic particles called neutrinos will be found to travel at just *under* the speed of light. However, please do not hold it against me if I turn out to be wrong on one or both counts!

Of the two announcements – the controversial news that some neutrinos can travel faster than the speed of light and the tentative discovery of the Higgs boson – the first one fits much more neatly into our definition of a scientific paradox.

The story so far is that a joint collaboration between two European labs, CERN in Switzerland and Gran Sasso in central Italy, has measured the speed at which a beam of neutrinos can cover the 454 miles between the two labs travelling underground through solid rock, which they can do as though it were empty space because they hardly ever interact with anything. In fact, billions of neutrinos, mainly produced in the Sun, are at this moment streaming through your body without your noticing.

At the centre of the OPERA collaboration (which stands for Oscillation Project with Emulsion-tRacking Apparatus) is a large, sophisticated instrument at Gran Sasso that can catch a tiny fraction of these elusive particles. In September 2011, its scientists announced they had clocked the neutrinos produced at CERN arriving sixty billionths of a second sooner than light could cover the same distance. That's not a lot faster: but it was still an incredible result.

For, according to our understanding of the laws of physics, nothing can exceed the speed of light. And in my experience, there is nothing that annoys people more about Einstein's theory of relativity (for that is where this notion originates) than its claim to this cosmic limit. Since Einstein's work in 1905, thousands of experiments have only confirmed it – and indeed, much of the beautiful edifice of modern physics rests on its being correct. The crucial point is not that light is particularly special, but rather that this speed limit is written into the fabric of space and time.

But what if Einstein was wrong? Is there a way of understanding the findings of OPERA? The whole point of a scientific theory is that it is there to be shot down – to be shown to be false by new experimental evidence or to be replaced with a better, more accurate theory that explains more. But extraordinary claims require extraordinary evidence, and the scientists working on OPERA – who cannot be faulted for the thoroughness of their experimental work – were the first to admit that they had no idea how their result could be possible.

After the media hype claiming Einstein was wrong came the next twist in the drama. A rival experiment at Gran Sasso, called ICARUS, also captured some of the CERN neutrinos, but measured their energy rather than their journey time. It had been pointed out by theorists very soon after OPERA's initial announcement that if the neutrinos were indeed superluminal (travelling faster than light) they would have to be emitting radiation throughout their journey and continuously losing energy. Not doing so would be a bit like an aircraft that manages to break the sound barrier without a sonic boom. It just shouldn't be possible.

The ICARUS researchers announced that they had found no evidence of this radiation, since the neutrinos arrived at their destination with the same energy they had left with. So the particles could not have been travelling faster than light.

The point is that ICARUS no more proves Einstein right than OPERA proves him wrong. Both results are experimental measurements, not discoveries. A proper test would involve a new experiment carried out independently by another laboratory. This is where, I believe, the speed of light will preserve its world record.

But I would love it if neutrinos could indeed travel faster than light. Such a discovery, if confirmed, would be heaven for physicists around the world. Blackboards would be scrawled on, heads scratched and Nobel Prizes tantalizingly within reach for a new Einstein able to solve the Paradox of the Neutrinos.

Index

Professor Jim Al-Khalili, OBE is an academic, author and broadcaster. He is a leading theoretical physicist based at the University of Surrey, where he teaches and carries out research in quantum mechanics. He has written a number of popular science books, translated so far into twenty languages, his most recent being *Pathfinders: The Golden Age of Arabic Science.* He has presented several television and radio documentaries, including the BAFTA-nominated *Chemistry: A Volatile History* and *The Secret Life of Chaos.* He presents the weekly science programme *The Life Scientific* on BBC Radio 4. He was awarded the 2007 Royal Society Michael Faraday medal and the 2011 Institute of Physics Kelvin medal, both for his science communication work.